PRENTICE-HALL BIOLOGICAL SCIENCE SERIES

William D. McElroy and Carl P. Swanson, *Editors*

SECOND EDITION

AN
INTRODUCTION
TO
ANIMAL
BEHAVIOR

Ethology's First Century

PETER H. KLOPFER

Duke University

PRENTICE-HALL INC., ENGLEWOOD CLIFFS, NEW JERSEY

Library of Congress Cataloging in Publication Data

KLOPFER, PETER H.
  An introduction to animal behavior.

  (Prentice-Hall biological science series)
  Bibliography: p.
  1. Animals, Habits and behavior of. 2. Animals,
Habits and behavior of—History. I. Title.
[DNLM: 1. Behavior, Animal. QL751 K66i 1974]
QL751.K593 1974   591.5   73-22467
ISBN   0-13-477935-5

**Revised from 1st edition
by P. H. Klopfer and J. P. Hailman**

© 1974, 1973 by PRENTICE-HALL, INC.
Englewood Cliffs, New Jersey

10 9 8 7 6 5 4 3 2 1

Printed in the United States of America

PRENTICE-HALL INTERNATIONAL, INC., *London*
PRENTICE-HALL OF AUSTRALIA, PTY. LTD., *Sydney*
PRENTICE-HALL OF CANADA, LTD., *Toronto*
PRENTICE-HALL OF INDIA PRIVATE LIMITED, *New Delhi*
PRENTICE-HALL OF JAPAN, INC., *Tokyo*

CONTENTS

PREFACE
TO
SECOND
EDITION

The purpose of this revision is unchanged from that originally stated in the first edition: a backbone text for an introduction to studies of animal behavior. This edition has taken into account both the advances in factual knowledge and the changes in theoretic outlook from 1950 to 1970. One additional resource should be mentioned to our readers. There is now available (in paperback form) a collection of those articles which, by and large, represent the development of ethology. *Function and Evolution of Behavior and Control and Development of Behavior* (edited by P. H. Klopfer and J. P. Hailman) provides a record of some important milestones in ethology, and is a useful supplement to this text.

This revision benefited greatly from the assistance and criticisms of numerous individuals, particularly the Duke Behavior Group of 1971–1972. Much of the library research and suggestions for organization for Chaps. 4, 6, and 7 were done by Lee McGeorge, Bill Hallahan, and Ray Barnett, to whom a special thank you is offered. I also wish to express my appreciation to Catherine Dewey and Sarah Freedman for their conscientious secretarial service. To Professor Fluke, Chairman of the Department of Zoology, I am indebted for providing that so-hard-to-define atmosphere which is a vital prerequisite to creative activities. My research has been supported by grants from the N.I.M.H, N.I.C.H.H.W, and a Research Scientist Award.

PETER H. KLOPFER

PREFACE
TO
FIRST
EDITION

It is our intention that this book serve as the "backbone" for a one-semester, senior or graduate-level course in animal behavior. But we also hope that the survey will introduce to that hypothetical creature, "the intelligent layman," the fascination of animal behavior studies. As with any skeleton, it requires flesh and blood to clothe the bare bones. This we expect to be provided by readings from the original sources and selected reviews, especially those listed in the annotated bibliographies following each chapter. A more comprehensive text than this could be written; however, we believe this volume should provide enough of an outline so that even the beginning student of animal behavior can use it to steer through the ocean of behavioral papers now flooding the scientific literature on all continents. We believe that a comprehensive text that allows less time for readings from original sources would be less profitable. In any event, this is how we have chosen to conduct our courses, and personal preference, after all, still remains the most compelling criterion for any book.

Ethology—the study of animal behavior from a biological viewpoint—is really a new science, although we reckon its beginning with the work of Charles Darwin. Now, a century after Darwin, ethology is expanding at an unprecedented rate, a huge endeavor which only the ambitious would attempt to summarize *in toto*. We have not attempted a synthesis of presently held explanations of behavior. Instead, we have tried to provide the novitiate with a summary of the past sweep of ideas and the experimental results of those men who founded and developed ethology.

In the chapters that follow we shall first attempt an historical survey of early work on animal behavior, dating from the middle of the nineteenth century, at the time of Darwin, until the end of that century (Part I and Chap. 1). We will then consider a more coherent movement, which, for lack of a better term, we will call *Classical* or *European Ethology* (Part II and Chap. 2). This "school" is characterized by a certain unity in the concepts and techniques employed by its adherents. This will be followed by a contrasting view of what we may term *American Ethology* or *Comparative Psychology* (Chap. 3). This includes works by a disparate group of men, largely psychologists, whose interests, while in some ways similar to those of their European counterparts, have led them down very different alleys. Chapters 4 through 8 deal with more specific problems that have occupied ethologists, from the role of hormones to animal orientation. Chapters 3 through 8, on disciplines related to ethology from 1900 to about 1950, comprise Part III of the book. Finally, in Part IV we have attempted a synthetic overview that explains animal behavior, not in terms of one school but in terms of all schools. This recent work (since 1950) has been fitted to our own view of the "causes and origins" of behavior.

Our chapter-by-chapter surveys in Parts II and III are intended to be relatively complete only through 1950, though later references have been included wherever this more recent work has *significantly* modified an earlier conclusion or where the later references represent reviews of earlier work. We have selected only what we consider the most striking departures from the pre-1950 studies for inclusion in our brief summaries of more recent developments. To those of our friends and colleagues who feel their work has been slighted, our apologies.

We should state that by behavior we mean those ways in which the organism adjusts to and interacts with its environment. To specify the definition further would only confuse matters, although this definition will certainly include phenomena not usually studied by ethologists (e.g., physiological functions such as the control of breathing rate by the concentration of carbon dioxide). The examples in the survey which follows constitute a class of phenomena whose characteristics define more or less what we mean by behavior as the province of ethologists.

Ethologists as students of behavior are charged with explaining the varied responses of animals to the different stimulus situations that they encounter. *Ethology* may be simply defined as the study of behavior from a biological viewpoint, a definition which connotes comparisons among species, as well as the physiological, ecological, and evolutionary aspects of the subject. Part IV attempts to present what we believe to be the underlying causal factors of behavior—the questions to which ethological research is ultimately directed.

Our view of the causes and origins of behavior differs, as will be seen, from the views of earlier ethologists. It differs, too, from the views currently held by some psychologists who distinguish between "molar" and "molecular" aspects of behavior. An example may help clarify their view: Watch a fly. Waltzing across your breakfast table, it probes the substrate, sampling the spilled remnants of your meal. When it encounters a pool of sugar or syrup, it firmly implants its proboscis and pumps the sweet matter into its gut. Should the fly stumble into a bit of pepper or salt, however, it rapidly retreats and moves off to seek yet another portion.

We could investigate the fly's organization through an analysis of the properties of its physical structure, i.e., use a *molecular* approach. We must then examine the sensory receptors by means of which the environment is sampled. We could discover how the receptors act and the course which the nerves follow to the central nervous system. Further, we could uncover the functions and *modus operandi* of the relevant constituents of the central nervous system itself.

As an alternative to the analysis of components of a behavioral system, we can attempt to understand the variables that affect the probability that an animal will make a given response to a particular stimulus. In the case of the fly, we might find that these variables include the duration of the prior period of food deprivation or its age. When all such variables and their influences and interactions have been considered, it is sometimes possible to infer certain principles of action or organization which underlie the overt behavior. Such principles or hypothetical mechanisms allow one to predict the response of the organism in a given state, though they would not necessarily tell us anything about the actual mechanics or design of the fly's nervous system.

We can in like manner describe the developmental changes that take place in these animals as they become older and, by studying flies of related and unrelated species, it may be possible to construct a "phylogeny" of behavior. Such a phylogeny can often tell us something about how complex acts evolve from simple acts, or, alternatively, how once complex acts degenerate into simpler ones. And, finally, still without considering the nature of the brains, nerves, and sense organs on which the behavior depends, we can concern ourselves with the adaptive function of the responses, thereby gaining an understanding of the evolutionary factors that lie behind the behavior in question. This entire class of explanations is what is meant by *molar* approaches to behavior.

The view which underlies our thinking, and which is detailed in Part IV, differs from the molar-molecular approach in small, but important, ways. We feel that there are really three separate molar causes (the selective advantage or function, the phyletic history, and the ontogeny of the behavior in question), and that the mechanisms of behavior must be studied in conjunction with the interactions of the organism with the environment in order to understand the full "control" of behavior. In the case of the fly, then, we should have to know something of all the causes listed above in order to achieve a complete understanding of its behavior. Thus, a really competent ethologist, in addition to a knowledge of the tools and techniques of his own trade, must know either a fair amount of genetics and evolutionary theory or neurophysiology. Ideally, he would know some of both. That he generally does not, represents a concession to the exigencies of time.

No scientific work is written in a vacuum. Throughout its preparation the book has benefited from the ideas and suggestions of Jeremy Hatch and the careful proofreading of Elizabeth Hailman. Drs. Colin Beer and Daniel S. Lehrman helpfully criticized certain chapters, while Dr. Dale Lott provided comments on Chap. 3, and Dr. Knut Schmidt-Nielson on Chap. 4. Dr. Vance Tucker kindly suffered through the entire manuscript. We are grateful to our academic bosses, Drs. E. C. Horn, J. Schwartzkopff, and D. S. Lehrman for the recognition that library research and pedagogic writing are as great a responsibility for a scientist as laboratory research and classroom teaching. During the time of preparation, while at Duke University and Rutgers, The State University, J. P. H. held pre- and postdoctoral fellowships from the National Institute of Mental Health; P. H. K.'s work was supported by the N.I.M.H., the National Science Foundation, and a Career Development Award from N.I.H. Mrs. Julie Berger and Mrs. Joyce Reiss carried through the many steps in the preparation of the manuscript, while our wives, Martha and Liz, were helpful and understanding throughout evenings of desertion. The death of the junior author's father just prior to completion of the final manuscript deprived the book of an educated and sensitive critic.

PETER H. KLOPFER
JACK P. HAILMAN

AN
INTRODUCTION
TO
ANIMAL
BEHAVIOR

Part I

FOUNDATIONS OF ETHOLOGY

1850-1900

Chapter 1

MENTAL EVOLUTION IN ANIMALS

A Chronologic Survey of Early Work

Introduction

The term *ethology*, coined some two centuries before Darwin, has had different meanings at different times. The seventeenth-century ethologist was an actor, a portrayer of human characters on the stage (see Jaynes, 1969, for the detailed history of ethology from which this summary stems). A century later John Stuart Mills redefined the term to mean the study of ethics (including the formation of character). Its use in this way continued into the twentieth century. In mid-nineteenth-century France, meantime, ethology became a label for "the study of animals, not as corpses reeking with formaldehyde in the Cuvierian tradition, but as living things in their natural habitat" (Jaynes, *ibid.*, p. 602). This sense of ethology apparently failed to become popular; not until the final quarter of the nineteenth century, largely through the writings of Alfred Giard, was ethology revived. Giard used the term to refer to studies that related animals to their immediate environment, a field which might today be labeled "aut-ecology." Giard distinguished ethology from comparative psychology, which entailed study of the behavior of individual animals. These terminological conflicts and confusions were further complicated by the idealistic stances of their antagonists. Giard was an ardent Lamarckian (which is reflected by his emphasis on the environment-organism relationship). As the Lamarckian position lost primacy, so also did Giard's use of ethology. Heinroth in Germany, W. Wheeler in the United States, and G. Bohn in France all continued to use ethology in this naturalistic sense, but stripped of any relationship to a Lamarckian position. Nonetheless, references to the term diminished, not to increase until the late 1940's. Its use then arose largely independently of the earlier precedents. Ethology came to designate naturalistic studies of animal behavior, in contrast to the excessively artifactual laboratory studies that had come to characterize comparative psychology. But this pertains to the history of the term. What of the studies of behavior themselves?

In Darwin's time ethology could scarcely be thought to have had an existence as a separate discipline. Ethology was rather an area in which any lover of animals (and especially of animal anecdotes!) could successfully dabble. As a body of ethological literature began to develop, and particularly in the twentieth century, as the relevance of Darwinian teachings to ethology became evident, specific schools and controversies that were directed to particular questions began to emerge. Thereupon, principles and theories tied together many disparate studies, allowing authors of textbooks

4

some basis for the organization of their material. But in 1850 this basis did not exist.

While evolutionary theories had, since the days of Democritus, made periodic appearances, it was not until the middle of the nineteenth century that evolution was acknowledged an inescapable fact of life. The effects on the biological sciences were galvanic. Ethology, so long concerned solely with problems of animal husbandry and pedagogy, instead focused on the problems of mental evolution (for a summary of pre-Darwinian ethology, cf. C. J. Warden, 1927). The comments and quotations on the following pages will reveal the existence of the following specific lines of inquiry:

1. Descriptions of the behavioral capacities and characteristics of organisms occupying different positions along the evolutionary continuum. From these accounts came inferences as to the evolutionary past of complex behavior. In some cases, the result of these inquiries was to set man apart from other beasts; in others, it was to emphasize his relationship.
2. Analysis of the laws of mental evolution. In general, these analyses concentrated on the relation between instinctive or innate behavior, on the one hand, and learned or acquired behavior on the other. Depending on the theological axe a given author was grinding, he emphasized the differences or the continuity between inherited and acquired behavior.

With the close of this century a new breed of ethologist arose. First among them was surely Lloyd Morgan, and then the "school" of Loeb. These men disdained the sterile controversies regarding the differences between the innate and the acquired (all too often "distinguished," it should be added, according to whether consciousness was present or not). Instead, they deliberately selected a methodology that precluded consideration of all factors not susceptible to experimental control. With Morgan, then, we see the start of an experimental ethology, whose course we shall trace in the following chapters.

To follow are summaries of the principal work of the early ethologists (in the modern sense of the term) from 1850 to 1900 in a more or less chronologic sequence. Such a simple chronology becomes useless after the turn of the century when both an increase in the number of ethologists and the development of coherent theories intertwined once separate studies. Excluded from this survey is the work of most psychologists whose contributions have been so insightfully considered by Boring (1957). Only when one of these men had something to say that was of particular and direct importance to ethology is he included here. Nor is every ethologist included. We selected those men whose work can fairly be said to have had an influence (whether for good or ill) beyond its immediate time. Finally, we had to contend with the fact that some of the men whose lives span both the nineteenth and twentieth centuries might be just as appropriately treated in a subsequent chapter. The assignment of Loeb and his followers to this section, for example, was quite arbitrary.

Survey

SPENCER: THE CONTINUITY OF MENTAL STATES

It is appropriate to begin our survey with Herbert Spencer, who, having been born in 1820, produced his major work shortly before the publication of Darwin's *The Origin of Species* (1859). Spencer was a thoroughgoing evolutionist. He anticipated a number of Darwinian notions, particularly those dealing with the inevitability of evolution, the continuity of mental states, and the development of habits and instincts. Unlike Darwin, he did not buttress his speculations with the mass of facts and detailed arguments which gave such force and cogency to *The Origin of Species*. Of greatest interest to us here is Spencer's *Principles of Psychology*, completed in 1855. It was followed some years later by additional volumes on first principles of biology, sociology, and morality.

Spencer's evolutionary doctrine rests on the assumption that any homogeneous mass when acted on by a force will not be altered throughout; consequently, a once homogeneous mass will become heterogeneous, the heterogeneity ever increasing with time. Spencer thus considers the transition from a homogeneous to a heterogeneous state a fundamental and universal process. This general notion was applied to specific cases as, for example, the development of a many-celled (heterogeneous) individual from an (homogeneous) egg. Many of the ideas for this notion were borrowed from the embryologist von Baer, who described the specializations successfully undergone by vertebrate embryos.

Although Spencer repudiated the notion of special creation, his concept of change was "orthogenetic." The end of a developmental chain was perhaps not predetermined, but nonetheless the evolutionary process, once initiated, proceeded mechanistically and inevitably. Of natural selection in the Darwinian sense he makes no suggestion. Indeed, Spencer later disagreed with the writings of Darwin and Wallace, believing that evolution was caused primarily by the inheritance of acquired characters; natural selection, Spencer insisted, was merely one of several subsidiary causes of biological evolution. In fact, he himself coined the phrase "survival of the fittest" to describe natural selection, but it was certainly not accorded the role of major or principal factor in evolution. For that matter, even the in-heritance of acquired characters did not represent an ultimate principle of evolution. In Spencer's view, the ultimate principle was to be found in the change from homogeneity to heterogeneity, which, in turn, is the conse-quence of yet more fundamental principles (discussed in his volume, *First Principles*).

Although Spencer's *Principles of Psychology* was published prior to *The Origin of Species*, it nonetheless clearly states the author's belief in con-tinuity among the forms of psychic life from the lowest to the highest. The lowest manifestation of psychic life he considered to be the reflex and the highest the volitional response. These grade imperceptibly into each other and a qualitative distinction between them and between organisms showing differing degrees of mental development cannot be found. Spencer (1896: 453–454) argued:

> That the commonly-assumed hiatus between Reason and Instinct has no existence, is implied both in the argument of the last few chapters and in that more general argument elaborated in the pre-ceding part. The General Synthesis, by showing that all intelligent action whatever is the effecting of correspondences between internal changes and external co-existences and sequences, and by showing that this continuous adjustment of inner to outer relations progresses in Space, in Time, in Speciality, in Generality, and in Complexity through insensible gradation, implied that the highest forms of psy-chical activity arise little by little out of the lowest, and cannot be definitely separated from them. Not only does the recently-enunciated doctrine, that the growth of intelligence is throughout determined by the repetition of experiences, involve the continuity of Reason with Instinct; but this continuity is involved in the previously-enunciated doctrine.
>
> The impossibility of establishing any line of demarkation between the two may be clearly demonstrated. If every instinctive action is an adjustment of inner relations to outer relations, and if every rational action is also an adjustment of inner relations to outer relations; then, any alleged distinction can have no other basis than some difference in the characters of the relations to which the adjustments are made. It must be that while in Instinct the correspondence is between inner and outer relations that are very simple or general; in Reason, the correspondence is between inner and outer relations that are complex, or special, or abstract, or infrequent. But the complexity, speciality, abstractness, and infrequency of relations, are entirely matters of degree. From a group of two co-existent attributes, up through groups of three, four, five, six, seven co-existent attributes, we may step by step ascend to such involved groups of co-existent attributes as are exhibited in a living body under a particular state of feeling, or under

a particular physical disorder. Between relations experienced every moment and relations experienced but once in a life, there are relations that occur with all degrees of commonness. How then can any particular phase of complexity or infrequency be fixed upon as that at which Instinct ends and Reason begins?

Even the distinction between psychical and physical phenomena is not absolute. In general, according to Spencer, psychic functions occur successively while physical functions may occur simultaneously. Also, as a consequence of evolution the psychic functions become increasingly localized in space, becoming concentrated, for example, within the cerebral hemispheres in higher mammalian forms.

Spencer provides little information on the actual details of the behavior of specific forms, descriptive details which are characteristic of so much of the later work in comparative psychology. Nonetheless, his emphasis on the continuity of psychic evolution marks him as an evolutionary-minded comparative psychologist who certainly influenced those who succeeded him.

DARWIN AND ROMANES: ETHOLOGY'S FOUNDATIONS

The three works of Darwin which have most directly influenced the development of theories of behavior are *Expression of the Emotions in Man and Animals* (1873), *Variation of Animals and Plants under Domestication* (1868), and *The Descent of Man* (1871). In the first of these books, Darwin develops three principles to account for the development of expressions and gestures: (1) the principle of serviceable, associative habits; (2) the principle of antithesis, and (3) the principle of the direct action of the excited nervous system on the body (all explained below). In his volume on variation, he details certain changes in behavior that are a consequence of selection. He traces such changes in order to provide the basis for a behaviorally based phylogeny. In *The Descent of Man*, finally, he discusses "displays" and "sign stimuli," subjects that assume great importance to the European ethologists of the next century (cf. Chap. 2).

Actually, most of these topics were treated in earlier works. The subject of displays and releasers, for instance, was alluded to in the course of his discussion of *sexual selection* (in *The Origin of Species*, 1859: 89): "Through natural selection in relations to different habits of life" the sexes may come to differ. Thereupon, the females (or males) have a basis for mate selection, the degree to which one individual possesses all the attributes of his (or her) sex being variable. Darwin goes on to suggest that the characters involved in sexual selection (e.g., elaborate plumage, movements, manes, excessive antlers, etc.) may have no other selective value. Ethologists today are less inclined to attribute differences in sexual characteristics to the esthetics of mate selection. Rather, such characters are adjudged important (and derive their selective advantage) because of their signal function. Nonetheless, the notion of sexual selection cannot be wholly divorced from our modern views on sexual displays and releasers (cf. Chap. 2). The careful reader will find countless other allusions in *The Origin of Species* which represent a portent of theories to come.

First, let us summarize the three "principles" listed above. We shall then briefly consider the relation between behavior and the processes of inheritance.

1. The principle of serviceable associated habits:

> Certain complex actions are of direct or indirect service under certain states of the mind, in order to relieve or gratify certain sensations, desires, etc.; and whenever the same state of mind is induced, however feebly, there is a tendency through the force of habit and association for the same movements to be performed, though they may not then be of the least use. Some actions ordinarily associated through habit with certain states of the mind may be partially repressed through the will, and in such cases the muscles which are least under the separate control of the will are the most liable still to act, causing movements which we recognize as expressive. In certain other cases the checking of the one habitual movement requires other slight movements; and these are likewise expressive. (Darwin, 1873: 28.)

In other words, when the particular stimuli which evoke "certain states of the mind" have repeatedly been associated with other stimuli that are responsible for the elicitation of some act, the performance of that act will eventually be elicited reflexly by the former stimulus alone or by stimuli similar to it. If a distinction between this process and the conditional reflex (Chap. 3) of modern psychology cannot be made, it is no wonder!

On the basis of extensive observations of animal behavior, Darwin had recognized one of the fundamental types of learning. But, Darwin believed the primary importance of such "associated habits" lay in the fact that they could be inherited. For this no convincing evidence is at hand. The anecdotal accounts Darwin relied on are open to alternative interpretations, and experiments specifically directed to the question (cf. McDougall, 1938) have been equally unsatisfactory. Indeed, in the light of our present knowledge of genetics it appears impossible that a phenotypic change could induce a specific genotypic change such as to allow for the inheritance of the acquired trait. The only exceptions one might expect would be found among microorganisms whose cells are not differentiated into somatic and germinal tissues. It is possible, of course, for mutations to occur whose effects mimic those of the acquired trait. If the trait in question possesses a selective advantage, the mutant may spread, thereby creating the impression of a Lamarckian inheritance. This phenomenon has been termed the *Baldwin Effect* and is often cited to explain the situations that would otherwise, in defiance of our knowledge of genetics, be claimed to prove Lamarck correct (cf. Simpson, 1953).

2. *The principle of antithesis:*

> Certain states of the mind lead to certain habitual actions, which are of service, as under our first principle. Now when a directly opposite state of mind is induced, there is a strong and involuntary tendency to the performance of movements of a directly opposite nature, though these are of no use; and such movements are in some cases highly expressive. (Darwin, 1873: 28.)

The validity of this principle rests on the dual assumption that (1) the statements relating to the principle of associated habits are true and (2) that it is possible to recognize movements of a directly opposite nature. Figure 1–1 does, indeed, suggest that the two stances shown are opposites. However, this interpretation is at least partly a matter of human judgment. That the two postures shown have different effects on other dogs can be demonstrated, although Darwin's phrase "though these are of no use" suggests that he did not appreciate the social value of such signals (cf. Chap. 2 on releasers). Whether different movements are due to neural mechanisms that are mutually incompatible, as is implied by the second assumption, can in principle be determined. At the time Darwin wrote, however, there was no theoretical and only scant empirical justification for assuming directly opposing or mutually inhibitory neural processes or centers.

3. *The principle of the direct action of the excited nervous system of the body, independently of the will and in part of habit:*

> When the sensorium is strongly excited, nerve force is generated in excess, and is transmitted in certain definite directions, depending on the connection of the nerve-cells, and partly on habit: or the supply of nerve-force may, as it appears, be interrupted. Effects are thus produced which we recognize as expressive. This third principle may, for the sake of brevity, be called that of the direct action of the nervous system. (Darwin, 1873: 29.)

Darwin believed that actions derived from the strong stimulation of pre-existing nerve chains generally combined with those that followed from the actions of the first or second principle. Therefore, most emotional re-

Fig. 1–1. (a) Dog approaching another dog with hostile intentions. (b) The same in a humble and affectionate frame of mind. (From Darwin, 1873.)

sponses represent a complex configuration. This point will emerge in a strikingly similar form when we consider the evolution of releasers and displays (Chap. 2).

In a revised edition of his volume on domestication (1868: 369–370) Darwin presented his theory of *pangenesis.*

> It is universally admitted that the cells or units of the body increase by self-division or proliferation, retaining the same nature, and that they ultimately become converted into the various tissues and substances of the body. But besides this means of increase I assume that the units throw off minute granules which are dispersed throughout the whole system; that these, when supplied with proper nutriment, multiply by self-division, and are ultimately developed into units like those from which they were originally derived. These granules may be called gemmules. They are collected from all parts of the system to constitute the sexual elements, and their development in the next generation forms a new being; but they are likewise capable of transmission in a dormant state to future generations and may then be developed. Their development depends on their union with other partially developed or nascent cells which precede them in the regular course of growth. Why I use the term union, will be seen when we discuss the direct action of pollen on the tissues of the mother-plant. Gemmules are supposed to be thrown off by every unit, not only during the adult state, but during each stage of development of every organism; but not necessarily during the continued existence of the same unit. Lastly, I assume that the gemmules in their dormant state have a mutual affinity for each other, leading to their aggregation into buds or into the sexual elements. Hence, it is not the reproductive organs or buds which generate new organisms, but the units of which each individual is composed. These assumptions constitute the provisional hypothesis which I have called Pangenesis.

Darwin went on to argue that not only do morphologic alterations affect the distribution and abundance of certain gemmules, but so do mental changes. Thus, a habit imparted to members of the parental generation may be directly transmitted to the offspring. In this Lamarckian view, Darwin found support both in the writings of his predecessors (cf. Spencer, for an example) and current public beliefs.

If Charles Darwin can be said to have been the first to attempt a rigorous and systematic study of comparative animal behavior and its evolution, George John Romanes may quite properly be called his first student. Romanes was actually born rather late in Darwin's life, 1848, and as he died in 1894, his life spanned only a portion of that century which is connected with the name of Charles Darwin. Yet he, more than any other biologist of his time, is responsible for placing the study of animal behavior on an evolutionary and truly comparative basis.

His first and most important early study was reported in the book *Mental Evolution in Animals* (1884). In this volume he attempted to provide a comparative analysis of mental function and its evolution. This was followed by his *Mental Evolution in Man* (1889) in which he concentrated on the development of intelligence and argued for the essential similarity of reasoning processes in man and the higher animals. In both of these books he included unpublished material from Darwin's manuscripts which Darwin, pleased that someone else was prepared to apply his notions of natural selection to problems of animal behavior, had given him. Darwin could scarcely have entrusted his material to a more sympathetic biologist, for Romanes proved one of his most enthusiastic supporters and disciples.

The ultimate aim of comparative psychology, according to Romanes, is classification, just as Darwin believed in classification as one of biology's ultimate aims. This was not classification in the classical manner, however, but a classification of psychological traits that revealed phyletic affinities and allowed one to trace the course of mental evolution. This was accomplished by careful analysis of the capabilities or responses of different kinds of animals with an eye to noting similarities and differences. Romanes saw the evolutionary course as a progressive development of discrimination ability and an enhancement of the power of adapted response. Consciousness, too, Romanes believed developed gradually in phylogeny, its presence being indicated by ever-increased response latency.

Romanes devoted considerable time and effort to the question of what consciousness is, under what conditions it occurs, and when it can be seen to be present. Where consciousness can be shown present, one may infer that a mind exists, but this begs the question. How can we show consciousness to be present? The form of this question may easily lead one to a solipsist position, the position that one can never know what occurs in other minds. But, Romanes, who never wanted to be bothered by philosophic uncertainties, suggested that the presence of consciousness can be known "ejectively." Ejective knowledge is knowledge of another's mental state that is inferred. Such inference is possible, due to the similarity in response patterns between ourselves and others when we are all confronted with the same stimuli. When we see a bear, we are aware of a conscious state that this leads to within us, and we are aware of the response that follows this conscious state. When we see another person making a similar response to that same stimulus, the bear, we may then infer that a similar conscious state exists in that other person. This is an ejective inference. Romanes argued that ejective inference represents an acceptable basis for knowledge. In general, the ejective clues to the presence of mind and consciousness are adaptive actions appropriate to particular stimuli. To quote Romanes (1884: 18): "The distinctive element of mind is consciousness, the test of consciousness is the presence of choice, and the evidence of choice is the antecedent uncertainty of adjustive action between two or more alternatives." This is not a very rigid definition, but it is interesting that it provides an objective measure for the presence of mind, this being the ability to discriminate between stimuli and to make adaptive responses which show some evidence of choice, i.e., they are variable. Further, Romanes argued that the physical basis of mind must in some way lie within the central nervous system with mental processes being the psychic equivalent of metabolic processes. Direct anatomical examinations of central nervous systems and their attendant parts can thus allow the drawing of inferences as to the degree of development of mind in the organism in question.

The mind itself, Romanes reasoned, is hierarchically organized into various faculties. These include reflex, instinct, emotions, reasoning, judgment, and volition, and they represent different points along a continuum, one having arisen gradually from its predecessor. Hard and fast lines cannot be drawn between them, Romanes stated. Instinct, for example, merely represents reflex action into which is imported the element of consciousness. As this element of consciousness is increased emotions may occur, and these, in turn, may lead to reasonable action. The contrast of these views with those of certain ethological schools (to be discussed later) is of some interest, particularly since the notion of continua is, if anything, more sophisticated and presents more subtle difficulties than the notion of discrete categories. Spencer saw instincts merely as compound reflexes with no consciousness involved at all. Spencer's consciousness, when it appeared, appeared suddenly, abruptly, and implied a rather discrete difference in neural organization.

With his strong evolutionary bias, Romanes was greatly concerned with the mode of origin of the different elements in the mental hierarchy. It is of special interest to us here to consider his hypotheses respecting the origin of instincts.

Romanes argued that there are two kinds of *instincts*: these may be called *primary* and *secondary*. The first are the direct result of natural selection. That is, it may be advantageous for a particular response to appear in full the first time the appropriate stimulus is presented. It may happen that as the result of chance variation one individual of a population will make this response. Having, therefore, a selective advantage over his fellows, he will be more likely to leave survivors than they, and, in time, the characteristics of this individual (including the instinct in question) will dominate the population. Romanes, of course, writing well before R. A. Fisher and his genetical theory of natural selection, was not able to make any statements about gene mutations or the quantitative effects of different degrees of selection pressure. He was able to see, as clearly as Darwin, how primary instincts could arise as the result of the normal variation found in every species.

The secondary instincts represent a mode of origin which Darwin himself had postulated as very likely possible, but which we today consider less credible. If an act is frequently repeated, Romanes argued, it may lead to an inherent change. This is the principle of *lapsing intelligence*, and is often illustrated by a man's performing a habitual action which in time becomes automatic without any element of consciousness intruding itself into the act. Romanes, along with Darwin, suggested that such habits, which might in time come to be unconsciously formed, could in some unknown manner come to be directly inherited and thereby result in *lapsing intelligence* or the formation of habits and instinct. The apparent contradiction with his earlier notion of instinct as a reflex action into which there is imported the element of consciousness did not seem to bother Romanes. Presumably, the element of consciousness involved in such a repetitive, habitual action is sufficiently small as to be unnoticed on most occasions, but it is still present in sufficient amounts to allow for the original distinction between instinct and reflex. In any event, as the differences between reflexes and instincts are considered to be continuous rather than discrete, no major difficulties arise.

The notion of lapsed intelligence, of course, is a thoroughly Lamarckian position and generally discredited today. Despite this, the modernity of certain of Romanes' views, particularly those respecting the relationship between reflex and instinct, mark him as a man whose outlook would be wholly compatible with many of those of his present-day colleagues.

It is evident from the foregoing accounts that most of the questions concerning behavior have addressed themselves to the evolutionary relation between volitional and automatic responses. First of all, considerable effort was devoted to the framing of a proper distinction between these latter two categories of response; then attention was focused on their manner of origin. Ingenious as were the results, the naiveté with which Darwin accepted the accounts of other owners or observers of animals is surprising. Also surprising is Darwin's failure to deduce at least some of Mendel's principles. For instance, the blending theory of inheritance in which Darwin believed, implies a loss of 50 percent per generation of the variability shown by any character. For variability to be maintained (or increased under domestication, as Darwin noted) it is necessary to assume an extremely high mutation rate. A particulate theory of inheritance avoids the difficulty altogether, for with such a system the parental store of variance can be maintained undiminished through any number of generations (Fisher, 1930). Quite possibly the acceptance of a particulate theory would have encouraged a more rigorous and quantitative study of inheritance and a more rapid demise of Darwin's theory of pangenesis. It was this last notion, it will be recalled, which provided an explanation, and thus a justification, for the inheritance of acquired characters, including habits.

MORGAN: HIS CANON AND LAMARCKIAN INHERITANCE

Lloyd Morgan's most important contribution undoubtedly was his refutation of the widely held Lamarckian views of his contemporaries. We can best summarize Morgan's argument by means of the following extract from his *Habit and Instinct* (1896: 319–321):

1. In addition to what is congenitally definite in structure or mode of response, an organism inherits a certain amount of innate plasticity.
2. Natural selection secures
   (a) Such congenital definiteness as is advantageous.
   (b) Such innate plasticity as is advantageous.
3. Both a and b are commonly present; but uniform conditions tend to emphasize the former; variable conditions, the latter.
4. The organism is subject to
   (a) Variation, of germinal origin.
   (b) Modification, of environmental origin, affecting the soma or body tissues.
5. Transmissionists contend that acquired somatic modification in a given direction in one generation is transmitted to the reproductive cells to constitute a source of germinal variation in the same direction in the next generation.
6. It is here suggested that persistent modification through many generations, though not transmitted to the germ, nevertheless affords the opportunity for germinal variation of like nature.
7. Under constant conditions of life, though variations in many directions are occurring in the organisms which have reached harmonious adjustment to the environment, yet natural selection eliminates all those which are disadvantageous, and thus represses all variations within narrow limits.
8. Suppose, however, that a group of plastic organisms is placed under new conditions.
9. Those whose innate plasticity is equal to the occasion are modified, and survive. Those whose plasticity is not equal to the occasion are eliminated.
10. Such modification takes place generation after generation, but, as such, is not inherited. There is no transmission of the effects of modification to the germinal substance.
11. But variations in the same direction as the modifications are now no longer repressed and are allowed full scope.
12. Any congenital variations antagonistic in direction to these mod-

ifications will tend to thwart them and to render the organism in which they occur liable to elimination.

13. Any congenital variations similar in direction to these modifications will tend to support them and to favour the organism in which they occur.

14. Thus will arise a congenital predisposition to the modifications in question.

15. The longer this process continues, the more marked will be the predisposition, and the greater the tendency of the congenital variations to conform in all respects to the persistent plastic modifications; while

16. The plasticity still continuing, the modifications become yet further adaptive.

17. Thus plastic modification leads, and germinal variation follows; the one paves the way for the other.

18. Natural selection will tend to foster variability in given advantageous lines when once initiated; for (a) the constant elimination of variations leads to the survival of the relatively invariable; but (b) the perpetuation of variations in any given direction leads to the survival of the variable in that direction. Lamarckian palaeontologists are apt to overlook this fact that natural selection produces determinate variation.

19. The transmissionist fixing his attention first on the modification, and secondly on the fact that organic effects similar to those produced by the modification gradually become congenitally stereotyped, assumes that the modification *as such* is inherited.

20. It is here suggested that the modification *as such* is not inherited, but is the condition under which congenital variations are favoured and given time to get a hold on the organism, and are thus enabled by degrees to reach the fully adaptive level.

When we remember that plastic modification and germinal variation have been working together all along the line of organic evolution to reach the common goal of adaptation, it is difficult to believe that they have all along been wholly independent of each other. If the direct dependence advocated by the transmissionist be rejected, perhaps the indirect dependence here suggested may be found worth considering.

As did Romanes, Morgan concerned himself with the problem of the origin of instincts and consciousness. In the case of the latter, he displayed a cautiousness that has been apotheosized as *Morgan's Canon*: never assume a response to be due to the action of consciousness when an explanation of the response is possible that avoids that assumption. Sentience may be possessed by the fertilized egg, Morgan argued, but as long as development proceeds in a predetermined course, whatever "consciousness" the egg possesses has no functional effect. It is only when conscious states influence the subsequent course of behavior that they assume any significance.

How, then, does consciousness arise? Assume an instinctive motor co-ordination that results in an integrated act, e.g., the pecking response of a newly hatched domestic chick. The motor response produces internal afferent impulses (the "back-stroke" of Morgan); these impulses may then produce the phenomenon of consciousness. This "back-stroke" may subsequently modify the instinctive motor act. Thus, if the chick's first (instinctive) peck is directed toward an evil-tasting insect, the sensations produced by the latter will assure that, upon the next encounter, the impulse to peck will be stayed or re-directed to a less obnoxious target. Such is the manner of development of "instinct-habits" and the explanation for the fact that most behavior is compounded of instinctive and volitional (or learned) elements.

Let us leave generalities, and take a particular case, fixing our attention on the very first occasion on which a chick pecks instinctively at a grain of food or other such object at suitable distance. The logical possibilities are as follows:

1. The action may be completely automatic.
2. The action may not be completely automatic, but in some degree *guided* by consciousness.

If completely automatic, it may be either—

A. *Accompanied* by consciousness.
B. Unaccompanied by consciousness. . . .

Let us assume, to begin with, that the first possibility expresses the facts of the case, and that the chick is, so far as the activity in question is concerned, a completely unconscious automation. It may be conceded that the first peck can be quite adequately explained on this view; response following stimulus under conditions which are purely organic and wholly within the sphere of the merely automatic co-ordination of the lower centres. It may, indeed, be urged that it is not reasonable to suppose that the chick pecks at the grain without seeing it, and this implies the presence of consciousness. But in the first place, we are not in a position to affirm that there is anything more than a physiological stimulus of the retina; and, in the second place, even if we were, it would still be conceivable that, though the incoming effects of the retinal stimulus give rise to the sight of the grain, the outgoing nerve-currents which call the muscles into co-ordinated activity (and it is this co-ordinated response that constitutes the instinctive activity as such) are not only automatic, but unconscious. Indeed, there is much to be said in favour of the view that the outgoing nerve-currents, and the molecular processes in the lower nerve-centres to which they are proximately due, have no conscious accompaniments in the case of the first performance of an instinctive act. Still, looking at the instinctive response as a whole and in all its bearings, there seem good reasons for supposing that, if it be automatic, it at any rate in some way affords data to consciousness; and that if the chick be an automaton, it is not merely an unconscious automaton. But how, it may be asked, can the observer decide whether consciousness be thus present as an accompaniment, though, so far, without guiding influence? Only by watching the subsequent behaviour of the bird. We are forced, then, to consider the after-effects in order to determine whether the first peck is accompanied by consciousness or not. And if we do thus watch the bird in its subsequent efforts, we find that they rapidly improve in accuracy, and soon have all the appearance of being under guidance and control, so that they can be modified or checked according to the nature of the object, nice or nasty, as the case may be. Now, we may safely lay down this canon: *That which is outside experience can afford no data for the conscious guidance of future behaviour.* When we say that conduct is modified in the light of experience, we mean that the consciousness of what happened, say yesterday, helps us to avoid similar consequences to-day. If the happenings of yesterday were unconscious, they could afford no data for to-day's behaviour. If, then, the first peck is unconscious, it is as such completely outside the experience of the chick, and can therefore afford no data for subsequent guidance control. Similarly, the second, third, and succeeding pecks, so far as automatic and unconscious, afford no data to experience. But observation shows that the activities concerned in pecking are not only guided to further perfection, but play a part in that active life of the little bird which cannot without extravagance be interpreted as unconscious; for only by appealing to consciousness can they be thus guided. Hence we seem forced to reject the hypothesis of unconscious automatism on the grounds that the activities in question do afford data to experience, can be modified, and are therefore subject to voluntary control, by giving rise to sensations and feelings which enter into the conscious life of the chick.

Let us assume, then, in accordance with the second possibility, that the very first peck is carried out under the guidance of consciousness. Now, guidance and control are based on previous experience. A chick, for example, will seize the first soldier-beetle he meets with; but after one or two trials of this distasteful morsel, though he may run towards one, if he catches sight of it moving at some distance from him, he checks himself so soon as he sees clearly what it is. He controls his tendency to peck at it in the light of his previous experience of its unpleasant taste. It is clear, however, that the first time a chick pecks there is no individual experience in the light of which the activity can be guided or controlled. Hence we can only admit the second possibility on the hypothesis that experience is inherited. But though it is quite conceivable that the effects wrought by experience are transmitted in some way, at present unexplained, through heredity —that, for example, the acquired skill of one generation may become congenital in the next—this is something very different from the inheritance of experience itself. Unless we are prepared to admit some form of metempsychosis; unless we believe that the individual remembers that which happened to its parents or grandparents;—we must hold fast to the fact that the conscious experience of the individual is limited to the events of its own lifetime. Remembering, then, that the phrase "inherited experience" is merely a condensed expression for the hereditary organic effects, if such there be, wrought through experience, we are forced to conclude that in the case of the first peck the chick has no experience in the light of which its action could be guided and controlled. If consciousness be present it is not yet effective. And our second hypothesis is thus placed out of court.

We are, therefore, thrown back upon our third possibility—that the automatic response gives rise to consciousness in the light of which the chick's future activities may be guided and controlled. On this view the first peck—and it must be remembered that we are concentrating our attention on the very first occurrence of an instinctive response in the course of individual life—though it is an organic and automatic response, nevertheless affords data to consciousness; it thus provides the initial experience by which subsequent efforts at pecking may be guided and controlled. (Morgan, 1896: pp. 129–133.)

LINDSAY: THE MORALS OF LOWER ANIMALS

Each generation and every discipline can claim some pedants. The strongest claim to that title, at least of the nineteenth-century ethologists, can be made on behalf of W. L. Lindsay, a self-styled physician-naturalist. His widely read *Mind in the Lower Animals* consists largely of lists of attributes, of definitions, or of undocumented anecdotes and assertions concerning the character of various beasts. Yet, because Lindsay illustrates so vividly the methodology against which Lloyd Morgan (and others) railed, it is worthwhile to consider him.

Lindsay's basic premise is that the psychic differences between men and animals are differences in degree only. To prove this, he lists the psychic attributes of infants and savages, contrasting these with various animals. Man often comes off second best! Indeed, so strangely informed is Lindsay that his "data" (anecdotal accounts, the truthfulness of which he claims to have personally established) suggest a closer genetic affinity between Englishmen and dogs, than, say, between Englishmen and Eskimos (the latter being "*beasts of prey*, without any other pleasure than eating. . . . [The

Eskimo] devours as long as he can and as much as he can get, like the vulture or the tiger. . . . He eats only to sleep, and sleeps only as soon as possible to eat again.' " [Lindsay, 1880: 122]).

> Dogs or other animals that may be considered in their way civil-
> ised or humanised—both as regards the individual and the race or
> breed—that have been subjected to persistent and judicious training
> by man—exhibit a manifest superiority to whole races or classes of
> man, both civilised and savage, in the following respects, which in-
> clude the *noblest of the human virtues*:—
>
>   1. Heroism, patriotism, self-sacrifice.
>   2. Compassion or sympathy, charity, benevolence, forgiveness.
>   3. Love and adoration of a master.
>   4. Fidelity to trust, duty, or friendship.
>   etc.

(Lindsay, 1880: 119).

Will, purpose, choices, feelings, all are exhibited by protozoa. As for insects . . . !

> Among the general mental characteristics of the *Insecta* as a class
> are the following:
>
>   1. Great variations of temper and disposition.
>   2. Likes and dislikes.
>   3. The passions, feelings, or emotions of fear, anger, or rage,
>   love, sorrow, impatience, pleasure and pain.
>   4. Appreciation of beauty in form, colour, and sound, including
>   musical tones and call notes.
>   5. Ingenuity or fertility of resource in difficulty, including the
>   use of tactics and stratagems in procuring food.
>   6. Acquisition and application of knowledge gained from experi-
>   ence.
>   7. Reception and communication of information, including the
>   exchange of ideas.
>   8. Formation of associations for specific objects—mutual assis-
>   tance, society, or emigration.
>   9. Obedience to orders.
>   10. Making deviations from routine in constructive or other op-
>   erations.

(Lindsay, 1880: 61).

Lindsay orders all vertebrates into a series of decreasing intelligence and psychic sensitivity. Presumably, it was this sort of thinking that lay behind Great Britain's vivisection ordinances, which include a prohibition on experiments with animals occupying a particular position in a hierarchical series if a "lower" animal would do instead.

The instinct-reason dichotomy attracted Lindsay's attention, too. After enumerating the "popular" and "scientific" definitions of instinct, he asserts a Lamarckian view: "All instinct is . . . inherited experience." (op. cit., 129). In this, of course, he had a clear precedent in Spencer.

A great part of Lindsay's *Mind in the Lower Animals* is devoted to an inquiry into the social, moral, and religious sensibilities of animals. The discourse gives little credit to Rousseau's noble savage, nor does it enlighten the reader interested in animal behavior. Yet, Lindsay's work is not wholly devoid of merit: he provides (op. cit., 32–34) a rather detailed list of contemporary ethologists and the nature of their interests.

WILLIAM JAMES: FOUNDATIONS OF COMPARATIVE PSYCHOLOGY

William James, it has been said, was as good a novelist as his brother Henry was a psychologist. Yet, to the ethologist, James's *Principles of Psychology* (1890) offers far more than a pleasing literary style. In the first place, his broad ranging survey of the problems of perception and learning are of applicability and relevance to animals other than man. Secondly, he explicitly anticipated present-day notions on the mechanisms whereby changes in species-characteristic behavior may occur. The terms "imprinting" and "critical period" were not known to him, but he accurately described the processes to which they refer.

To begin with, James saw instincts as reflexes, and the nervous systems of most organisms as bundles of co-ordinated reflexes. These were established through evolution by natural selection. James, in fact, specifically disclaims the notions of Lamarck (and Darwin) concerning the acquisition of instincts and supports Weismann's "disproof" of Lamarckianism (based on the fact that in metazoans the genetic tissue is segregated from the somatic tissue that responds to environmental stimuli).

These instincts, or reflex bundles, need not be invariable, however (James, 1890: 391, footnote):

> In the instincts of mammals, and even of lower creatures, the uniformity and infallibility which, a generation ago, were considered as essential characters do not exist. The minuter study of recent years has found continuity, transition, variation, and mistake wherever it has looked for them, and decided that what is called an instinct is usually only a tendency to act in a way of which the *average* is pretty constant, but which need not be mathematically "true."

Two principles govern variations in instincts. It is these which we now recognize in the phenomena of imprinting and critical periods (cf. Chap. 2). James (1890: 394–395) terms the first of these principles "inhibition by habit":

> The law of inhibition of instincts by habits is this: When objects of a certain class elicit from an animal a certain sort of reaction, it often happens that the animal becomes partial to the first specimen of the class on which it has reacted, and will not afterward react on any other specimen.
>
> The selection of a particular hole to live in, of a particular mate, of a particular feeding-ground, a particular variety of diet, a particular anything, in short, out of a possible multitude, is a very widespread tendency among animals, even those low down in the scale. The limpet will return to the same sticking-place in its rock, and the lobster to its favorite nook on the sea-bottom. The rabbit will deposit its dung in the same corner; the bird makes its nest on the same bough. But each of these preferences carries with it an insensibility to *other* opportunities and occasions—an insensibility which can only be described physiologically as an inhibition of new impulses by the habit of old ones already formed.

He supports these comments with an account of the experiments by Douglas Spalding, a contemporary ethologist with an empirical bent (see below).

James's second principle, which we recognize as a statement concerning critical periods, is named the "law of transitoriness" (James, 1890: 398–399, Vol. II):

> *Many instincts ripen at a certain age and then fade away.* A consequence of this law is that if, during the time of such an instinct's vivacity, objects adequate to arouse it are met with, a *habit* of acting on them is formed, which remains when the original instinct has passed away; but that if no such objects are met with, then no habit will be formed; and, later on in life, when the animal meets the objects, he will altogether fail to react, as at the earlier epoch he would instinctively have done.
>
> No doubt such a law is restricted. Some instincts are far less transient than others—those connected with feeding and "self-preservation" may hardly be transient at all, and some, after fading out for a time, recur as strong as ever, e.g., the instincts of pairing and rearing young. The law, however, though not absolute, is certainly very widespread, and a few examples will illustrate just what it means.
>
> In the chickens and calves above mentioned, it is obvious that the

instinct to follow and become attached fades out after a few days, and that the instinct of flight then takes its place, the conduct of the creature toward man being decided by the formation or non-formation of a certain habit during those days. The transiency of the chicken's instinct to follow is also proved by its conduct toward the hen. Mr. Spalding kept some chickens shut up till they were comparatively old, and, speaking of these, he says:

"A chicken that has not heard the call of the mother until eight or ten days old then hears it as if it heard it not. I regret to find that on this point my notes are not so full as I could wish, or as they might have been. There is, however, an account of one chicken that could not be returned to the mother when ten days old. The hen followed it, and tried to entice it in every way; still, it continually left her and ran to the house or to any person of whom it caught sight. This it persisted in doing, though beaten back with a small branch dozens of times, and, indeed, cruelly maltreated. It was also placed under the mother at night, but it again left her in the morning."

The instinct of sucking is ripe in all mammals at birth, and leads to that habit of taking the breast which, in the human infant, may be prolonged by daily exercise long beyond its usual term of a year or a year and a half. But the instinct itself is transient, in the sense that if, for any reason, the child be fed by spoon during the first few days of its life and not put to the breast, it may be no easy matter after that to make it suck at all. So of calves. If their mother die, or be dry, or refuse to let them suck for a day or two, so that they are fed by hand, it becomes hard to get them to suck at all when a new nurse is provided. The ease with which sucking creatures are weaned, by simply breaking the habit and giving them food in a new way, shows that the instinct, purely as such, must be entirely extinct.

The sum total of James's contribution to psychology need not be considered here. Readers may turn to the excellent account by Boring (1957), among others, for a detailed exposition and evaluation. Yet, we could not exclude James from our account of animal behavior studies for his prestige and his work provided a necessary impetus that helped ethology free itself from Lamarckian doctrines.

FABRE, FOREL, AND THE PECKHAMS: INSECT BEHAVIOR

Insects have provided material for collectors for centuries, and no less for "collectors" of behavior. And of such collectors there has rarely been a shortage. (In nineteenth-century England it would appear that any aspiring clergyman must also be a competent coleopterist!) Both the earliest and, recently, the most exhaustive studies of animal behavior have focused on insects. Perhaps the four best known of the nineteenth-century entomologists were J. H. Fabre (*The Wonders of Instinct: Chapters on the Psychology of Insects*, 1920), A. H. Forel (*Das Sinnesleben der Insekten*, 1910), and G. and E. Peckham (*On the Instincts and Habits of the Solitary Wasps*, 1898). Among them, these workers nicely illustrate the major positions exciting ethological passions of the last century.

Fabre, whose romantic descriptions of insect instincts have seldom been surpassed, objected vehemently to the notion that complex, precisely integrated instincts could be built up gradually through the stepwise acquisition and inheritance of component acts. Only the perfected instinct has survival value, Fabre argued, and unless it can have so arisen natural selection could never have promoted it. The complexity of many instincts, on the other hand, argues against their *de novo* appearance, except, presumably, under the aegis of Special Creation. Fabre thus rejected not only the Lamarckian explanation for the evolution of instincts, but explanations based on the selection of chance variations as well.

> If, on her side, the Wasp excels in her art, it is because she is endowed not only with tools, but also with the knack of using them. And this gift is original, perfect from the outset: the past has added nothing to it, the future will add nothing to it. As it was, so it is and will be. If you see in it nought but an acquired habit, which heredity hands down and improves, at least explain to us why man, who represents the highest stage in the evolution of your primitive plasma, is deprived of the like privilege. A paltry insect bequeaths its skill to its offspring; and man cannot. What an immense advantage it would be to humanity if we were less liable to see the worker succeeded by the idler, the man of talent by the idiot! Ah, why has not protoplasm, evolving by its own energy from one being into another, reserved until it came to us a little of that wonderful power which it has bestowed so lavishly upon the insects! The answer is that apparently, in this world, cellular evolution is not everything.
>
> For these among many other reasons, I reject the modern theory of instinct. I see in it no more than an ingenious game in which the armchair naturalist, the man who shapes the world according to his whim, is able to take delight, but in which the observer, the man grappling with reality, fails to find a serious explanation of anything whatsoever that he sees. In my own surroundings, I notice that those who are most positive in the matter of these difficult questions are those who have seen the least. If they have seen nothing at all, they go to the length of rashness. The others, the timid ones, know more or less what they are talking about. And is it not the same outside my modest environment? (Fabre, 1916: 377–378.)

How strikingly different is the prose of the Peckhams, and not least in the conclusions which it states:

The general impression that remains with us as a result of our study of these activities is that their complexity and perfection have been greatly overestimated. We have found them in all stages of development and are convinced that they have passed through many degrees, from the simple to the complex, by the action of natural selection. Indeed, we find in them beautiful examples of the survival of the fittest. (Peckham and Peckham, 1898: 236.)

The Peckhams, recall, were studying members of the same class of animals as Fabre! Their procedure was far more methodical, however. They classified the behavior of wasps into intelligent and instinctive acts, these two kinds of acts being more or less operationally distinguished. Instinctive acts were characters of a species and appeared full-blown on the first occasion they were evoked. Intelligent acts might vary from individual to individual and generally improved with practice. However, all intermediate grades were recognized, and the role of consciousness was studiously ignored.

Forel (1904), finally, represents a reversion to earlier views. His comments were primarily directed against the mechanistic views of Bethe (cf. discussion in Boring, 1957, and next section). Forel asserts an anthropomorphic position, endowing his ants with the consciousness and the sensations and perceptions (though to a lesser degree) of man.

LOEB AND THE MECHANISTS

Toward the close of the nineteenth-century there arose a "school" of ethology which asserted the errors of anthropomorphic and teleological reasoning. To Jacques Loeb is attributed the founding of this school with his *Der Heliotropismus der Tiere und seine Uebereinstimmung mit dem Heliotropismus der Pflanzen* (1890). (*Also*, cf. *Forced Movements, Tropisms and Animal Conduct*, 1918.) This was followed by a series of publications by a group of men who embraced Loeb's mechanistic doctrines: T. Beer, A. Bethe, and J. von Uexküll. About von Uexküll we shall have more to say in another context (cf. Chap. 6).

Loeb's study of plant movements impressed upon him the possibility of oriented movement in the absence of any organs of consciousness or sensation. He then proceeded to apply the principles of forced movements, or tropisms, to animals and discovered that, given enough tropisms, he could account for a wide range of activities without having to make either anthropomorphic or teleological assumptions (cf. Chap. 8).

The various tropisms by themselves, however, could not account for much of instinctive behavior, particularly when the behavior in question (e.g., breeding activities) appeared only at certain times of the year. This problem was surmounted by assuming that either the thresholds for the initiation of a tropistic response or the sign of the response (whether toward or away from the stimulus source) could be altered by certain chemicals (e.g., hormones). In this manner instinctive behavior could be fully integrated into a grand tropistic theory.

Once one has begun enumerating tropisms so as to satisfy all cases, the problem of where to stop can become an embarrassment. With no cogent reason to stop with reflexes or instinctive behavior, the mechanists proceeded to extend their domain over learned behavior as well. This was accomplished by arguing that memory images—which represent physical traces in a specific region of the brain and could thus not exist in forms lacking this organ—may also exercise an orienting influence.

> The tropistic effects of memory images and the modification and inhibition of tropisms by memory images make the number of possible reactions so great that prediction becomes almost impossible and it is this impossibility chiefly which gives rise to the doctrine of free will. The theory of free will orignated and is held not among physicists but among verbalists. We have shown that an organism goes where its legs carry it and that the direction of the motion is forced upon the organism. When the orienting force is obvious to us, the motion appears as being willed or instinctive; the latter generally when all individuals act alike, machine fashion, the former when different individuals act differently. When a swarm of *Daphnia* is sensitized with CO_2 they all rush to the source of light. This is a machine-like action, and many will be willing to admit that it is a forced movement or an instinctive reaction. After the CO_2 has evaporated the animals become indifferent to light, and while formerly they had only one degree of freedom of motion they now can move in any direction. In this case the motions appear to be spontaneous or free, since we are not in a position to state why *Daphnia a* moves to the right and *Daphnia b* to the left, etc. As a matter of fact, the motion of each individual is again determined by something but we do not know what it is. The persistent courtship of a human male for a definite individual female may appear as an example of persistent will, yet it is a complicated tropism in which sex hormones and definite memory images are the determining factors.
>
> Our conception of the existence of "free-will" in human beings rests on the fact that our knowledge is often not sufficiently complete to account for the orienting forces, especially when we carry out a "premeditated" act, or when we carry out an act which gives us pain or may lead to our destruction, and our incomplete knowledge is due to the sheer endless number of possible combinations and mutual inhibitions of the orienting effects of individual memory images. (Loeb, 1918: 171–172.)

It is easy to see why these doctrines provoked outcries from workers unwilling to cede man's unique position, or committed to a belief in free will, or, in animals, consciousness and purpose. It is also easy to see how the accuracy with which Loeb could measure responses and define stimulus parameters impressed the experimentalists of his time. What is less easy to fathom is Loeb's failure to recognize when his "explanations" became no more than the verbiage which he disdained in his antagonists.

OTHERS

During its earliest years, ethology was not a discipline that seemed to require any particular kind of professional training or acumen. Anyone with an eye for the antics of animals and the ability—once more generously fostered than today—to describe these literately (and, preferably, humorously) could style himself a contributor to this field.

Lubbock

John Lubbock (1899) was primarily known for his anthropological studies (horrifying as this may appear to the anthropologically sophisticated of today). In the study cited above, however, he turned to the problems of sensory physiology. What do animals perceive, Lubbock asked, and it is to his credit that he went beyond conjecture in an effort to obtain an empirical answer. The crudity of his experiments, the absence of controls, and the simplest statistical techniques should not blind us to the values of the example Lubbock set in attempting to secure answers through experiments.

Galton

Francis Galton's work (1899) is well enough known in other contexts to be disregarded here, though not until we emphasize the important role played by Galton's pronouncements on the nature of inheritance. His impressive studies of inheritance largely discounted the importance (to heredity) of acquired characters.

Spalding

Douglas Spalding was cited in the extracts from William James's study. However, it was not until the 1950's that Spalding was "rediscovered" (cf. Haldane's brief sketch, 1954, and Gray, 1962). The first experimental study of the following-response and the critical period in birds must be credited to him.

Overview

Before the end of the nineteenth-century, the concept of evolution by natural selection had become widely accepted. Ethology was then focused upon the processes and progress of mental evolution. First and foremost among the workers in this field were Spencer and Darwin, both of whom explicitly recognized a continuity in mental states and behavioral capacities from the lowest mammals through man. Both men also concerned themselves with the laws governing mental evolution. In particular, Darwin's principles of antithesis and associated habits have proven viable up to the present day. Darwin's lead was followed by Romanes who established the formal study of comparative behavior and set the stage for the first of the "modern" students of animal behavior, Lloyd Morgan. Morgan's caveats on the avoidance of superfluous suppositions played an important role in the development of objective and operational procedures in the study of behavior. The contributions of James were also vital to behaviorists struggling to free themselves from some of the misleading implications of Lamarckianism. Finally, Loeb and his mechanistic school grafted the methodology of the physical sciences onto the speculative root of ethology.

Selected Readings

References to the specific works of the men cited will be found in the bibliography at the end of this volume. Historical surveys of more general content are to be found in the articles by C. J. Warden (1927), "The Historical Development of Comparative Psychology," *Psych. Rev.*, Vol. 34, in the book by E. G. Boring (1957), *A History of Experimental Psychology*, and a history of the term "ethology" in J. Jaynes's (1969) article in *Animal Behavior*, Vol. 17.

Part II

THE STRUCTURE OF ETHOLOGY

1900-1970

Chapter 2

THE
STUDY
OF
INSTINCT

The Tradition of Lorenz and Tinbergen

Introduction

Studies by European ethologists such as K. Z. Lorenz and Niko Tinbergen have had a profound effect upon current research in animal behavior. The reason for this effect is threefold. First, almost all of the concepts of these ethologists stemmed directly from observation of the animal in its natural environment. The problems suggested by the observations were therefore both real and important. Second, the European ethologists were interested in causal factors of behavior at all levels: molar and molecular. Lastly, these men studied a wide variety of organisms, both vertebrate and invertebrate. This broad point of view toward animal behavior generated the first relatively complete conceptual framework for ethology. Therefore, it is appropriate that European ethology be considered before other, more disparate or specialized, schools of thought.

The efforts and results of the European ethological school culminated in the publication of Tinbergen's summary, *The Study of Instinct*, in 1951. The following survey is a development of concepts up to the time of Tinbergen's book. A new approach, synthesizing these notions with those of American comparative psychology and neurophysiology, began emerging in the 1950's. These later developments are discussed in the final chapters of this book.

European ethologists were naturalists. Their first objectives were always to watch the animal in its natural surroundings and to describe its behavior. Tinbergen grew up as an avid naturalist, watching insects and birds with the same excited curiosity that infects readers of his popular book, *Curious Naturalists* (1958). Lorenz and others preferred to rear animals under near-naturalistic conditions in order to observe their behavior closely. Lorenz's delightful *King Solomon's Ring* (1952) and *Man Meets Dog* (1954) are classics of popular animal lore. Whatever the exact nature of their focus, European ethological studies are permeated with an empathic, albeit not necessarily anthropomorphic, view of animal behavior.

Experiments concerning molar causes of behavior were usually "natural experiments" in which nature provided the variables. For instance, one might observe how an animal responds to its conspecifics or to a predator in the wild. When European ethologists performed more conventional experiments, the experiments were usually simple and were carried out in the field. For instance Tinbergen and Perdeck (1950) held simple cardboard models of the parent gull in front of chicks to find out what visual stimuli elicited the latter's begging. The nervous system was viewed as a "black box," the important properties of which were deduced from the environmental inputs and their relationship to the behavioral outputs.

The epithet "European" when applied to these ethologists refers merely to the geographic locus of their naturalistic studies. Most of these ethologists were in fact continental or English: Kortlandt, Armstrong, Baerends, Heinroth, Kirkman, Huxley, Lorenz, and Tinbergen to name a few; Nice, Craig, Whitman, and others worked in America. The school as a whole, however, was little influenced by comparative psychology during most of its development, and comparative psychology was a largely American enterprise. Because of this separation, the verbal dichotomy of European versus American has arisen as a convenient shorthand for distinguishing naturalistic ethologists from other groups of workers. When the Europeans finally crossed paths with one of the American groups, much controversy arose over the concepts of the European school (for instance, Lehrman, 1953; Schneirla, 1956). The nature of these disagreements, many of which have not been completely resolved, will be considered later (cf. Chaps. 3 and 12). Suffice it to point out here that European naturalists studied relatively stereotyped behavior patterns of (primarily) birds, fish, and insects in their natural environments, whereas their American critics studied the variable behavior of mammals under laboratory conditions. It seems likely that any two groups of men, equally matched in intellectual capacity and motivation, studying in these two divergent conditions would invariably evolve different explanatory concepts for animal behavior. As is the case with any progressing branch of science, the original approaches of both the European ethologists and their American critics are now somewhat dated. But we can still learn much from their ideas, methods, and results.

A convenient introduction to European ethology can be made by following the steps that led ethologists to their concepts. First, let us be naturalists, observing behavior, organizing its components, and searching for its meaning in relation to the environment. Then we shall concentrate on individual behavior patterns, describing them in detail from both the stimulus and response points of view. By comparing such descriptions among numbers of species, we will seek to deduce the phylogeny of behavior. We shall also see how descriptions of behavior give rise to assumptions about the nervous system and how it works. Finally, we shall consider the ontogenetic development of behavior.

Behavior and Its Natural Function

THE ETHOGRAM

Behavioral dossiers for many European animals were carefully compiled after long periods, sometimes years, of field observation. For instance,

Kortlandt (1940) described many activities of the European cormorant including such seemingly insignificant traits as head scratching, stretching movements, and "drying of the wings" as well as feeding, nest building, and courtship activities (see Fig. 2–1). Likewise, Makkink (1936) noted in detail all of the behavioral patterns of the avocet which he could distinguish in the field. Many such listings, called *ethograms*, have now been made for a variety of creatures.

The ethogram is the starting point of investigation by European naturalists. Tinbergen (1951: 7–8) made clear at the outset of his classical summary that "special emphasis should be placed on the importance of a complete inventory of the behavioral patterns of a species." The reason for such an emphasis is clear: premature experimental investigation may fail to take into account important factors which influence the behavior pattern under consideration. For instance, Adams (1962) points out that "many otherwise ingenious experiments have been largely vitiated by a failure to appreciate, for example, that a rat in a strange environment is positively thigmotactic."

NATURAL SELECTION AND BEHAVIOR

The naturalistically inclined ethologists classified, as well as named and listed, the behavior patterns they observed. A simple and natural classification could be based on the presumed function of the behavior, such as feeding, reproduction, escape from predators, and so on. This approach, in turn, led to specific investigation of the function of behavior. For instance, Makkink's (1936) study of avocets revealed that these birds often seem to sleep at peculiar moments, such as during the lull in a territorial fight. This led to the investigation of "displacement" or "out-of-context" activities (see pp. 46–47).

Since they approach ethology with interests in evolution and ecology instead of physiology, the Europeans quickly realized that the function of a behavior pattern could be expressed in terms of its selective advantage. For instance, many species evolved elaborate series of courtship interactions between mates because individuals having these patterns presumably left a greater number of offspring than those lacking them. However, the dynamics of selection were rarely investigated, and it is only recently that the survival value of enigmatic behavior patterns has been attacked experimentally. For instance, Tinbergen and his co-workers (1962) made an experimental analysis of why parent birds remove egg shells from the nest just after the chicks hatch (see Chap. 9).

COMMUNICATION AS A FUNCTION

Interest in the function of a behavior pattern in the animal's life led naturally to a return to the analysis of animal communication (Darwin's "expression of emotions," see Chap. 1). Early European ethologists at-

Fig. 2–1. Displays of the cormorant. (After Kortlandt, 1940.)

tempted to find the exact meaning of particular postures, movements, and vocalizations shown by animals to their conspecifics. For instance, Huxley (1914) was fascinated by the extraordinary postures given by courting grebes (Fig. 2–2); other workers described the movements of stickleback fish prior to spawning (Tinbergen, 1951), the postures of herring gulls (Tinbergen, 1953b), or the ruffling of the mane and the depressing of the ears in wolves (Schenkel, 1948). Each activity was classified according to the part of the life cycle in which it appears, such as courtship, care of the young, and so on. Eliot Howard (1920) wrote, "we can study life and we can study environment but we must not divorce them if we are to study behavior."

While studying great crested grebes, Huxley realized that postures in courtship had some function other than merely "expressing emotion." Specifically, emotion had to be expressed *to* something, and thus there arose the question "Why do animals communicate?" Huxley was unhappy about Darwin's conception of "sexual selection" (see Chap. 1) since it assumed that the male's courtship antics were evolved to satisfy some aesthetic sense of the female—which sense, for some reason, demanded vivid colors and active motions. (Other formulations of the phenomena posed equally difficult problems.) So Huxley suggested that courtship acted as a mutual excitant that led to earlier (therefore presumably more successful) breeding. This concept he called "mutual sexual selection."

Actually, Huxley's formulation stood somewhere between the original concept of sexual selection and the now accepted explanations developed by Pycraft, Tinbergen, and others. Specifically, close observations by ethologists showed that courtship movements and vocalizations serve many purposes, such as bringing mates together, stimulating hormonal flow necessary for reproductive activity, and ensuring synchrony of the male and female in the breeding cycle. A behavior pattern whose primary biological function is communication is called a *display*.

The study of displays became one of the chief interests of European ethologists. The title of one of the more influential books of this period, Armstrong's *Bird Display and Behaviour* (1947), symbolizes the strong emphasis both on displays and on the creatures whose displays were studied. With the clarification of the point that displays possess communicative functions, a re-interpretation of Darwin's "principle of antithesis" emerged. The expressions of "fear" and "anger" are not opposite merely because of internal restrictions of the animal's nervous system, as Darwin suggested (cf. Chap. 1), but rather because the expressions are meant to communicate to the companion entirely opposite "intentions" (fleeing versus attack). If two displays signifying opposite messages are as different (to the perceiving companion) as possible, confusion in meaning is greatly reduced. In the accounts that follow, bear in mind this emphasis on displays, for many concepts of the European ethologists developed from the analysis of displays.

The Unit of Stereotyped Behavior

Careful description of individual behavioral patterns showed that certain motor acts are nearly always produced only in certain biological situations. Stimulus and response are apparently linked in a constant and predictable manner.

THE MOTOR PATTERN

The *fixed action pattern*, as the motor act is often called, was seen as extremely stereotyped within a species, particularly in acts of communication. The motor pattern is alike in all members of the species, at least all members of the same sex, and alike in every repetition by an individual. These fixed action patterns (also called fixed motor patterns, *Erbkoordinationen* and *Instinkthandlungen*) form the core of behavioral investigation by the European ethologists.

Fixed action patterns, although stereotyped, may be continually redirected or adjusted to the location of signals of objects in the environment. An analogy would be the pulling of a rifle's trigger and the subsequent discharge (the fixed action pattern) as opposed to the pointing and aiming of the gun. The latter, steering component of the fixed action pattern has been termed *taxis* (Tinbergen, 1951; Lorenz, 1950). This is a different sense of taxis than that conventionally used, as by Fraenkel and Gunn (1940) (and note Moltz, 1965).

THE STIMULUS AND PERCEPTION

Each fixed action pattern tends to be elicited by relatively simple stimulus-objects in the environment. In fact, many such stimuli are the fixed action patterns of companions, which are said to "release" the observed behavior. Thus some actions and bodily structures were termed *releasers*, a term introduced by Lorenz (1935). The sensory pattern that the releaser creates was thought to correspond to a very specific receiving center in the perceiving animal's central nervous system. This receiving center is called the *innate releasing mechanism* (IRM). Lorenz uses the analogy of a key (releaser) fitting a lock (releasing mechanism).

The concept of the innate releasing mechanism implied that it (1) was innate, (2) responded to configural stimuli, and (3) was located centrally in the nervous system. Although we cannot deny that some animal perception might be organized in this way, no cogent example has been found. For instance, both Schleidt (1961) and Hailman (1964a) studied classical

(1)

(2)

(3)

(4)

(5)

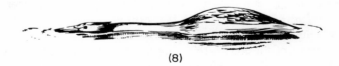

(6)

(7)

(8)

(9)

(10)

(11)

(12)

(13)

Fig. 2–2. Courting habits of the great crested grebe, *Podiceps cristatus*. (After Huxley, 1914.) (See also facing page.)

examples of the IRM and found that although a simpler, nonconfigurational perception was present before specific experience, a configural perception was only developed through learning (habituation in turkeys and conditioning in gull chicks; cf., Chap. 3). The gull chick's perception is probably organized peripherally, perhaps being modified by experience in more central parts of the nervous system. Furthermore, recent electrophysiological evidence (see Chap. 6) shows that some processing and selecting of information go on at all stages of sensory systems. It seems likely that the perceptual mechanism for responding to releasers may be organized in various ways in different animals and at different stages of ontogeny (see also Chap. 11).

Two famous examples of releasers were discovered in birds. David Lack (1943) showed that territorial fighting by the orange-breasted adult European robin was released by the orange breast of its neighboring rival. A stuffed juvenile, which lacked the color, did not evoke fighting, but a distinctly unbird-like tuft of orange feathers did! G. K. Noble (1936), whose principal contributions to ethology dealt with hormonal studies (see Chap. 5), made a similar experimental discovery in the colorful American woodpecker, the flicker. Males are distinguished from females merely by possession of a small black moustache. Noble painted a moustache on the female of a pair and noted violent attacks by the male upon his spouse. When the female was recaptured and the moustache removed, the male once again addressed her sexually. The results of this vivid and ingenious experiment were further confirmed with stuffed models presented near the nest.

Some environmental stimuli are not strictly releasers, since they are not structures, movements, sounds, or odors of a companion. Nevertheless, these stimuli—of which an example might be the specific shape and color of a flower that releases the approach of a bee—act on innate releasing mechanisms in much the same way as do releasers. The term *sign stimulus* became the name for this wider range of releasing stimuli, and the releaser might be considered a special type of sign stimulus. (Similar concepts go by many names: *auslösendes Schema* and *angeborenes Auslösende Mechanism* of Lorenz, trigger, *Symbolbewegungen* of Heinroth, innate perceptory pattern, and so on.)

Tinbergen and others noted that sometimes the innate releasing mechanism could be fooled into being unlocked by a stimulus similar to, but not identical with, the biologically correct sign stimulus or releaser. An oystercatcher will attempt to roll into its nest objects other than its own displaced egg, such as a gigantic model of an egg. Presumably this happens because the releasing mechanism is relatively simple, being only specific enough to correctly discriminate objects in the actual life of the animal that might be confused with the biologically meaningful object. Often such an artificial stimulus can be constructed that is even more effective than the normal one in bringing about the fixed action pattern. This object is said to be a *super-normal stimulus*.

DRIVE

An important part of this stimulus-response behavioral unit is missing from the concept as so far developed: the notion of *drive*. Craig (1918) had noted that animals sought out the releasing stimulus, as if internally driven to do so. He called this initial behavior *appetitive* because he imagined it to be evoked by an internal appetite for the required stimulus. Craig's "appeted stimulus," an earlier, unrefined version of the sign stimulus, brought about the fixed action pattern, in his terms, *the consummatory act*. After performing the act, the animal rested.

Notions of appetites or drives in the behavior studied by European ethologists help to define the behavior patterns as instincts. Unlike James (see Chap. 1), these ethologists did not view instincts as mere bundles of reflexes; rather, some intervening organization between stimulus and response characterized an instinct. The nature of drive, as conceived by these ethologists, is considered again below. In general, Lorenz and Tinbergen looked upon a specific consummatory *act* as itself the goal of a specific drive. Most ethologists today would adhere to a view more closely related to Craig's, namely that a specific stimulus *situation* is sought—a *consummatory stimulus*.

Observations on ducks, pigeons, grebes, and many other creatures led to this conception of instinctive behavior: (1) appetitive behavior or "unrest" leading to (2) the correct sign stimulus that triggers (3) a stereotyped motor act. Inferred from these observations were certain internal states (appetites), sensory organizations (innate releasing mechanisms), and other properties of the animal's nervous system, to be considered further below.

The concept of drive was generalized to various levels of organization (see discussion of hierarchical models, below) such that there existed not only drives or appetites for specific responses or goals, but also general drives that somehow contributed to the activation of the specific drives. For instance, the Groningen group under the influence of Baerends (e.g., Kruijt, 1964) believed there to be three major drives: attack, withdrawal, and sex. Displays were thought to be motivated by various combinations of these drives.

The entire concept of drive, in its various forms, was reviewed in a series of critical articles by Hinde (e.g., 1956, 1959, 1960, 1970). The essence of Hinde's objections is that drive has neither a unitary definition nor does it imply a unitary mechanism. At least six separate operational situations have been used to define and measure an underlying drive. Depending on the measurement chosen, the drive may be assessed at different strengths. The fact that such measurements of drive often correlate with each other did make the concept useful in the past. Hinde argues that today's accurate methods of assessing and predicting behavior, e.g., the sequential analysis of chains of behavior patterns (Chap. 11), renders a vague "drive" concept unnecessary. Finally, if drive is made operational (e.g., considered to be the probability of linkage between two behavior patterns), it follows that one cannot assume that drive energizes behavior in any way. "Drive" becomes merely a description of behavior without physiological implications. *Tendency* was suggested as an operational term to replace "drive," which has so many different meanings; however, "tendency" as an operational term was quickly prostituted, and it is now commonly used in a manner similar to the older concepts of drive (e.g., Kruijt, 1964).

Inferred Nervous Organization

THE HYDRAULIC MODEL

The abrupt execution of the consummatory act and the triggerlike action of the sign stimulus might call to mind a picture of behavioral "energy" inhibited by a block which is suddenly removed. In fact, Lorenz and others proposed pedagogical models of the behavioral unit.

One hydraulic model, affectionately known among behavior students as the "flush toilet," is shown in Fig. 2–3. In this model an endogenous drive builds up readiness to act by filling a reservoir with *action specific energy*. Presumably, as the reservoir fills up, the animal becomes increasingly restless and exhibits appetitive behavior. When the sign stimulus appears, signified by the weight, it pulls out the valve (innate releasing mechanism), allowing the energy to be channeled into specific pathways (the fixed motor acts).

The model accords admirably with behavioral facts. It embodies the concepts of internal driving force, the releaser, the releasing mechanism, and the consummatory act; but this is no surprise, since the model was conceived with the facts beforehand. The model also brings out another aspect of the behavioral unit thus far not considered: *intensity*. Sometimes fixed action patterns occur in different "strengths," movements forming continua of speed, completeness or amplitude of some motor act. These differences are embodied in the concept of intensity, which is shown to be part of the hydraulic model. Intensity is dependent on the amount of accumulated action-specific energy.

The model is decidedly aphysiological; it was never meant to imply

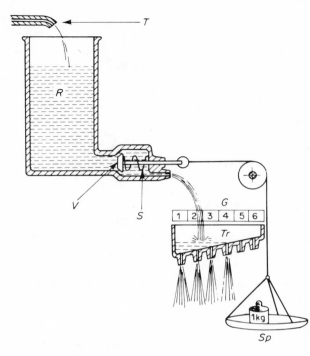

Fig. 2–3. Lorenz's hydraulic model of motivation. T, source of action-specific energy; R, reservoir of action-specific energy; V and S, components of the innate releasing mechanism; Sp, external releaser; Tr, motor pathways. The ease with which V and S can be operated to release R depends on the "heighth" of energy-fluid column R and the "weight" Sp. The particular motor pathway activated is similarly related to the heighth of energy-fluid in Tr. (After Lorenz, 1950.)

liquid reservoirs in the central nervous system. Rather, the model provided a convenient way of describing the general properties that the actual neural mechanisms must have. Thorpe (1956) and others have attempted to reinterpret the model in terms acceptable to current physiology. Reaction-specific energy becomes specific-action potentiality (SAP); instead of "energy" specific to effectors of the consummatory act, one can visualize specific neural channels to muscles. Drive is merely the endogenous firing by specific central nervous centers which control the act. Although this kind of re-interpretation merely makes the model a direct physiological hypothesis, actual neural mechanisms have yet to be discovered.

The revised model can be criticized on at least two important counts. First, the "energizing" drive concept has been shown by Hinde (see above,

p. 43 to be too vague to be useful as anything other than a very general impression of behavior. Secondly, as Lehrman (1953) and others have pointed out, the model contains no provision for proprioception (sensory feedback) in behavioral control. It is certainly probable that the rapid cessation of consummatory acts is in some cases due to interoceptive feedback from muscle spindles, rather than an exhaustion of neural energy in defined circuits.

DISPLACEMENT, VACUUM ACTIVITIES, AND HIERARCHICAL ORGANIZATION

Considerations of the hydraulic model and its variants were extended to explain other disparate observations on behavior. These have led to some of the most characteristic concepts of European ethology.

Incipient movements made by an animal (i.e., movements of very low intensity) were called *intention movements* (*Intentionsbewegung* of Heinroth). If a fixed action pattern were not released for a long period of time, what would become of the accumulated energy? Some ethologists suggested it somehow pushed through the releasing mechanism valve (Fig. 2–3) and escaped into the consummatory acts. Such an act given in absence of its usual releaser, in a behavioral vacuum as it were, is called a *vacuum activity* (*Leerlaufreaktion*). Or perhaps the energy could flow over the top of the reservoir and into some other fixed action pattern; in this case it is a *displacement activity*. More accurately, displacement activities seem to occur when a releaser is actually present, but the action pattern is somehow "thwarted" or in conflict with another action. It would be as if Fig. 2–3 had the holes of the action patterns plugged with something so that the liquid energy spilled over the top of the lower trough and into some other trough of action patterns.

The model and its variants were even further extended to the relationships between behavior patterns. Some patterns, such as certain courtship displays, tend to occur together. Groups of displays tended to occur together, such as courtship, nest building, incubation, and care of the young do in the reproduction of birds. From these observations a *hierarchical system* could be constructed, each level having a "flush toilet" whose action filled the reservoirs of several toilets on the next level.

Tinbergen (1951) did suggest such a model, which was more explicitly neuronal than Lorenz's hydraulic model (Fig. 2–4). Each level in Tinbergen's hierarchy is composed of "centers" that receive neural energy from internal and external sources (hormones, environmental stimuli, etc.). Each center passes the energy to subordinate centers when a neural "block" is removed by the innate releasing mechanism. Tinbergen attempted to couple this scheme in a hierarchical scheme proposed by Weiss (1941) to describe motor movements. For instance, a major instinct (e.g., reproduction) is

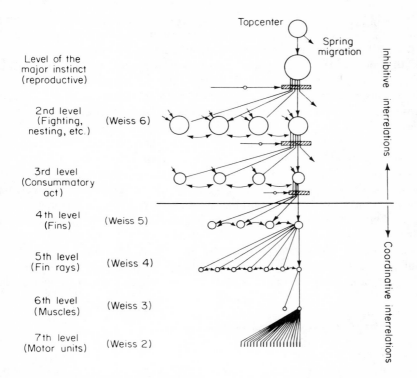

Fig. 2–4. The hierarchical system of "centers" underlying a major instinct, viz., the reproductive instinct of the male three-spined stickleback. (After Tinbergen, 1951.)

represented by a center that sends impulses to centers for fighting, nesting, and other activities. Each of these centers, in turn, sends energy to centers for specific motor patterns of, say, nesting. The motor pattern center then activates individual centers for appendages, for example, a fish fin, which in turn activates fin rays, then muscles in the individual fin rays, and finally the muscle units themselves.

Upon examination, it becomes difficult to synthesize this hierarchical model with other ethological notions. For instance, it is not clear just how appetitive behavior could be fitted consistently into this scheme. Nevertheless, most contemporary ethologists, although not ascribing to the details of such a hierarchical scheme, do admit that something like it must underlie the organization of behavior. (See also the comments under Recent Developments, below.) Kortlandt (1940) noted that fixed action patterns related in Tinbergen's hierarchy of drives were also related in the frequency with which one pattern was a displacement activity of another. (Kortlandt's model was electrical, with impulses "sparking over" from one pattern to another.) Kirkman (1937), after Adams (1931), simply called such behavior "substitution," thus avoiding physical analogues.

Finally, the action of hormones upon behavior was interpreted in terms of the model. Hormones were envisioned as acting centrally upon the higher levels in the hierarchy of control centers. Included here were also the effects of external stimuli which "motivated" behavior rather than releasing it. Internal stimuli were also thought to act directly upon these higher centers.

Some of these concepts which grew from the hydraulic model are now considered outdated. Intention movements (not envisioned as connoting conscious "intention" by the animals) and vacuum activities certainly do exist, but their physical bases are to be sought in neurophysiological, not purely behavioral, studies. Displacement appears to be a rubric covering many kinds of behavioral interaction which have different physical bases (Tinbergen, 1951; see also Chap. 11). And hormones clearly act in many ways upon the organism, peripherally as well as centrally (see Chap. 5). But the hydraulic model did serve a useful purpose in summarizing description of stereotyped displays and other behavior of a variety of animals.

Comparative Behavior and Evolution

Whitman and Craig reared many species of pigeons and doves for comparative studies; Heinroth and Lorenz raised ducks; Heinroth, a zookeeper, brought up many other European birds; and Tinbergen studied several species of gulls in the wild. What did they hope to learn?

Of course, species differences might give valuable clues as to how a behavior pattern is adapted to its environment. Or a dynamic process, difficult to study in one species, might prove more amenable to study in a similar species whose behavior requires a longer period of time. Many such reasons for comparative study exist. But it was in order to find clues to the phylogeny of behavior and of species themselves that many of the comparative behavior studies were undertaken.

THE EVOLUTON OF BEHAVIOR

The Germans, in particular, noted the analogy between a species-specific fixed motor pattern and an anatomical character. By the turn of the century it was accepted that comparative anatomy of a group often revealed a spectrum of structure and variation which showed that some species had "primitive" characteristics; for instance, the bones of a certain species in a group might resemble fossil bones of an ancestor of the group.

Comparative studies, in the absence of fossils, may provide a guess as to what the ancestor might have been like. In such reconstructions an assumption as to what characterizes "primitive" must be made. Usually, the least specialized or least complex characters are taken to be most like the ancestral ones; this is a dangerous assumption since characters can become more simple during evolution.

It was but a step further to studying fixed action patterns as if they were anatomical characters and to reconstructing the ancestor's behavior from a collection of the least specialized aspects of behavior found within a group. For instance, a long comparative study of gull displays beginning with studies by Goethe (1937) and others is still being carried out by Tinbergen (e.g., 1959) and his associates (cf. Chap. 10). Furthermore, species with a preponderance of unspecialized characters are assumed "primitive" and most like the ancestor. Groups of animals with similar behavior patterns are presumed to share a recent ancestor. In short, behavior was used as a taxonomic tool for the study of phylogeny and systematics. This taxonomic aspect, properly belonging more to systematics than to ethology, nevertheless represented a strong motive of many European ethologists and was given much impetus by Lorenz.

However, the evolution of particular behavior patterns, as opposed to the evolution of behaving organisms, could also be inferred from a comparative series. Two notions about behavioral evolution were in vogue before the turn of the century: (1) Lamarckian incorporation of acquired behavior into the gametes forming the next generation and (2) evolution of behavior by giant steps. The former was adhered to by Lindsay and to a lesser extent by Romanes while the latter notion was expounded by Fabre (1920) (see Chap. 1). Whitman (1919) was the first of many comparative ethologists to address himself to these questions.

Whitman clearly pointed out that rigidity of a conditioned reflex or other acquired behavior patterns did not prove that learned behavior is ancestral to instinctive behavior. In fact, he argued that much the reverse takes place in evolution, with unlearned behavior patterns giving way to those in which learned elements predominate. This continuum was somewhat foreshadowed by Romanes (1884), although he also allowed Lamarckian mechanisms (cf. Chap. 1). The wealth of evidence for the gradual evolution of behavior patterns through natural selection emerged from later studies, particularly of displays.

THE ORIGIN OF DISPLAYS

Many ethologists showed that displays could often have been evolved from noncommunicative behavior. Moreover, the ancestral pattern often persisted, either in the same species along with the display derivative, or in another species in a similar behavioral context. This evolution of displays from nondisplay behavior is called *ritualization* or emancipation. Huxley (1914) had understood this principle when he noted how a sexual display of the grebe could have been derived from a simple locomotory movement by which the bird approaches the edge of the nest (also, see Whitman, 1919).

It was just at the end of our survey period when Daanje (1950) showed that displays often were derived not from the full unritualized locomotor response but just from an incipient portion of it—an *intention movement* (see p. 38). These studies showed how natural selection could act to maintain and develop displays gradually, since the unritualized precursor as well as the evolving display are both useful in the animal's survival and reproduction.

Another possible origin of displays was suggested: displacement activities (see p. 46). An animal stimulated by the presence of a rival both to flee and to attack might perform some displacement activity which could through further evolution become a signal. Such signals in antagonistic situations would be of survival advantage since the rivals could assess the probabilities of their opponent's fighting at that moment or at that particular place. By displays two birds, for instance, could determine a mutual territorial boundary with a minimum of actual combat; the territorial "line" would emerge at that place where both individuals were equally likely to fight, this likeliness being signified by display.

Instinct and Imprinting

The similarity of both the stereotyped motor acts and the simple sign stimuli in related species suggested the inheritance of behavior. Craig, Whitman, Heinroth, and Lorenz noted that animals reared in isolation developed the correct motor acts of the species, even though the young animals never saw any conspecific perform them. Doves without experience went through courtship and bred successfully. Finally, many acts appeared suddenly, with no time for practice, and these acts were indistinguishable from similar acts of older animals who had performed them repeatedly. From these facts the word *instinct* was reborn.

Also called "innate behavior pattern" or "inherited behavior," it was the fixed action pattern that was central in the concept of instinct. Certainly, the term "innate releasing mechanism" implies a similar inheritance for the perceptual organization, but this perceptual aspect was more open to qualification (e.g., Lorenz's discussion in the *Kumpan* paper, 1935).

Stereotypy of behavior obviously proves neither the presence nor the absence of learning, however, and "instinct" became a target for criticism of European ethology (e.g., Lehrman, 1953; Schneirla, 1956). European ethologists performed only a few actual studies of the ontogeny of behavior, the only way in which the controversy could be resolved.

The most common type of ethological study of ontogeny involved rearing animals in isolation from their conspecifics (*Kaspar Hauser* animals) and seeing what behavioral patterns appeared. Such animals often showed the consummatory fixed motor patterns, although the patterns were often released by abnormal stimuli. Whitman's (1919) study of pigeons provides many examples of this phenomenon. Even these studies, however, were criticized by American psychologists (Lehrman, 1953; Schneirla, 1956), who held that certain motor patterns could be "self-learned" in isolation. For instance, a mother rat "learns" to lick and not to eat its newborn young because it licks its own genitalia prior to giving birth. The mother presumably becomes conditioned to licking in response to its own odor or taste, which permeates the young. A complete solution to behavioral ontogeny is still lacking (see Chap. 12), although the autonomous development of stereotyped motor patterns has now been observed in a multitude of species.

A fuller discussion of the ontogeny of stereotyped, species-specific behavior patterns is deferred to Chap. 12. Two final remarks should suffice here. First, some behavior patterns are exhibited in their correct context the first time they are exercised, when there has been no possibility for the usual kinds of learning described in Chap. 3. For instance, domestic canaries without previous experience with nesting material will immediately carry material to a nest site; they do not court it and they do not carry it to the middle of the cage floor (Hinde, 1958). This is not to say that their previous experience has not affected the development of the behavior in some respects. The second point, articulated by Spurway (1956), is that behavior *per se* is not inherited since it is the product of the interaction of the inherited genes and the environment; but two animals raised in the same environment may show different behavior patterns so that this *difference* can be spoken of as genetic.

In 1910 Heinroth announced an extraordinary discovery: young goslings follow the first relatively large moving object they see after hatching. The young geese subsequently follow the object in preference to anything else. Of course, in the wild, this object is usually the parent goose. Heinroth called this rapid fixing of social preferences "*Prägung*," the German term for impressing, as in stamping out a coin; Lorenz (1937) translated it *imprinting*. Lorenz further said that imprinting occurs rapidly, does not involve "rewards" as do usual kinds of learning (cf. Chap. 3), and lasts a lifetime. Geese reared by Lorenz preferred Lorenz to actual conspecifics at mating time!

Imprinting was known to William James (see Chap. 1), although it received no notice from his followers. Furthermore, Whitman (1919) had discovered a similar phenomenon before the turn of the century. The only way he could hybridize species of doves was to rear the young of one species in the nest of another. The dove, when grown, preferred individuals of its foster-parent species to actual conspecifics. However, these facts were only given publicity years later when, after his death, Whitman's book on pigeons appeared. Imprinting, in any event, represents a phenomenon which is in many ways a curious hybrid between the ethologists' instinctive and learned behavior (Klopfer, 1964; Bateson, 1966).

Some Recent Developments

Since 1970, European ethology has undergone three kinds of development: (1) corrections or refinements of certain concepts, such as the innate releasing mechanism and drive (mentioned above; and see Schleidt, 1962; Lorenz, 1965); (2) a change to more experimental and operational approaches to behavior, still based on firm observation of behavior in the natural environment; and (3) a broadening of the areas of interest and kinds of behavior studied. This broadening is in part a synthesis with other approaches to behavioral study (cf. Chap. 3). Because of the broad base of European ethology, nearly all of Part IV of this book could be said to be recent developments of this school.

In addition to the developments mentioned elsewhere there are a few recent contributions that we believe to be particularly important. First, the mechanisms for perception of stimuli that elicit fixed motor patterns can no longer be framed in a unitary concept such as the "innate releasing mechanism." Marler (1961b) reviewed the evidence and concluded that much stimulus filtering found in species-common behavior involves the whole sensory system as it functions in normal perception, although certain examples do imply special filtering mechanisms. The specificity of the latter filtering may be due to (1) receptor filtering; (2) filtering by the whole afferent pathway; and, possibly, (3) special mechanisms that correspond to the original IRM notion. Thus, as an example of receptor filtering, in certain butterflies for which the color red has particular value as a releaser, Bernhard et al. (1970) have shown the tapetum selectively to reflect that color onto the retina. Less specific "pathway filtering" is probably illustrated by the intromission "code" of certain rodents in which the reproductive responsiveness of the female depends on the male's employing a specific pelvic thrust frequency (Diamond, 1970). Finally, there do appear to be situations in which there is no apparent relation between the properties or maximal sensitivities of the receptors or neuronal detectors and the properties of the releaser. This suggests the existence of IRMs *sensu stricto*. Possibly the use of microelectrodes to record activity of single neurons in sensory systems will bring about a firmer basis on which to place speculations about the locus of sensory filtering (cf. Chaps. 6 and 11).

Aspects of the fixed motor patterns of species-common behavior have received less attention. Behavioral analyses are being made using increasingly smaller motor units as the basis for observation, as in the study of incubation and nest building in black-headed gulls (e.g., Beer, 1961, 1962, 1963*a*, 1963*b*). Further, the degree of stereotypy both among individuals of the same species and upon repeated performance of the same motor act is receiving increased attention. Both the frequency and duration of motor acts, as in duck courtship displays (Dane et al., 1959; Dane and van der Kloot, 1964), have been used to measure stereotypy. However, sophisticated computer analyses of a single display in the jackal have led Golani (1973) to reject altogether the notion of stereotypy in the motor components of a stereotyped display. The Gestalt remains stable while its elements vary. Further, it appears that a division between the "innate, fixed" part of a motor act and the "learned, variable taxis" is not a meaningful way to analyze all species-common behavior patterns (Hailman, 1964*a*).

Hinde's criticisms of unitary "drives" were mentioned earlier in the chapter (p. 44). The subject of drive is being approached in wholly new ways. Displacement activities, originally thought to be due to the conflict of drives, are now known to be behavior patterns of heterogeneous control or "causation" (see Chap. 11). The relationship between behavior patterns, and *changes in responsiveness to constant stimulation* are being studied in a more operational framework (e.g., Hinde, 1954). Andrew (1961) created a model to account for the mobbing calls by blackbirds given to predators, emphasizing the concept of a threshold of response. Nelson (1964) has approached the courtship of fish from a probabilistic viewpoint of temporal patterning (and see Slater, 1973). Male courtship could be described approximately as a first-order *Markov chain*; that is, each action could be predicted by knowledge of the immediately preceding action, but knowledge of prior actions does not increase the predictability. The female's actions, however, depended on the male's actions, and only upon the cumulative effect of the male's actions. Once a female performed a certain act, however, this act immediately altered the male's display sequences and types. This study also shows how closely the problem of communication is tied to the problem of motivation.

Daanje (1950) demonstrated that many displays derive evolutionarily from incipient locomotory movements. Morris (1957) further showed that displays can originate from autonomic responses, such as avian feather raising and lowering in response to temperature. Morris also articulated the concept of "typical intensity" in which a continuum of intensities of an act is broken up into discrete intensities for less ambiguous communication. Marler (1961a) analyzed displays from a logical basis, using communication concepts developed in linguistics, and Smith (1963) showed that the meaning of a particular display is not constant but depends on the behavioral context in which it occurs. The role of cybernetics in the analysis of communication will become increasingly prominent in future studies.

The effects of natural selection in maintaining a behavior pattern and the evolution of behavior patterns are discussed in Chaps. 9 and 10, respectively; few truly new advances have been made in these areas since 1950. The ontogeny of behavior is, however, beginning to yield many new explanations (see Chaps. 3 and 12). Imprinting is a major area of study; although it is too early to make a final summary of imprinting results, it does seem clear that imprinting is not a new, qualitatively different kind of learning; it may be, however, a unique combination of aspects from other kinds of learning. Most imprinting studies are directed to one or more of these questions (from Bateson, 1966):

1. What stimuli elicit initial approach of the young bird?
2. What stimulus configurations are actually imprinted?
3. What defines the critical or sensitive period in which the imprinting must take place to be effective?
4. How long-lasting and stable is the imprinted preference?

Song learning is another area of ontogenetic study that has developed since 1970 (see Chap. 12). Few studies have been made of motor development (cf., Hailman, 1964), but several have been made on sensory development (e.g., Schleidt, 1961; Hailman, 1964; Hinde, 1970; Nottebohm, 1970). Since ontogenetic studies have shown experience to be important in the development of species-common behavior, the study of "instinct" is no longer appropriate as the focus of European ethological study. Perhaps this school ought best to be thought of as studying naturally occurring, species-common behavior. Part IV summarizes many other recent developments of European ethology.

The original taxonomic bias of early ethology is being challenged anew partly in response to some overly facile extrapolations from birds or monkeys to men (e.g., Ardrey, 1970; see Klopfer, 1973). For instance, much human behavior is explained, justified, or excused on the assumption that it represents a biological heritage, ignoring the multiplicity of mechanisms that may serve common ends and *vice versa*. English robins defend individual territories; Galapagos mockingbirds have communal territories. Which species provides a better basis for explaining human behavior? Even if we confine our phylogenetic speculations to closely related groups, just the passerine birds, can an evolutionary sequence for territorial behavior be derived? Possibly, but only if those characteristics which resist evolutionary change can be distinguished from those which do change. Unfortunately, that distinction requires an *a priori* knowledge of the evolutionary sequence. In short, serious questions are being raised about the validity of the sort of evolutionary schemata represented by the study illustrated on p. 208 of the Empeid fly (Klopfer, 1973).

Finally, the concept of "*a* display," of the relation of the part (e.g., a muscle contraction) to the whole is once again beginning to excite attention. The central importance of the observer's original bias and his perceptual world, both as reflected in his language (Whorf, 1956) and his methodology are being critically re-examined (Nelson, 1973; Bateson, 1972).

Overview

European ethologists such as Tinbergen and Lorenz, in the tradition of Whitman, Craig, Heinroth, and Huxley, closely observed animals in their natural environment or in seminatural conditions of captivity. Their first object was to compile a complete behavioral dossier or *ethogram* of the species studied, and then to discover the function (evolutionary selective advantage) of each behavior pattern. A primary focus of attention was the *display*, a pattern evolved to serve a primarily communicative function.

A behavior pattern is elicited by a simple *sign stimulus* or social *releaser* stimulus which acts on a hypothetical central nervous system center, the *innate releasing mechanism*. The releaser triggers a stereotyped motor response or *consummatory act*. Often, an *appetite* drives the animal internally to seek the releasing stimuli so that variable *appetitive behavior* precedes the stereotyped act.

An hydraulic model summarized the dynamic characteristics of the behavior pattern, but it was taken literally by some critics to be a conception of the physiological mechanisms underlying the behavior pattern, which it was not. Rather, it was an analogy to aid the description of observed behavior.

Some stereotyped motor patterns are graded in *intensities* of speed, magnitude, or completeness of movement; incipient movements, particularly those of locomotion, are called *intention movements*. When a motor pattern is given without the presence of its usual releasing stimulus, it is said to be a *vacuum activity*. A term used to cover several different kinds of "out of context" behavior patterns is *displacement*. Patterns that substitute for one another in displacement appear also to be related functionally and form hierarchical organizations. These concepts were also part of the hydraulic model. More thorough physiological investigation of all these processes is under way.

Displays and other species-constant motor patterns can be used in taxonomic studies as if they were morphological characters. Comparative study further showed that natural selection can produce communicative behavior from nondisplay behavior, a process called *ritualization*.

Stereotyped patterns appear in animals reared in isolation, although they tend to be given in unusual stimulus situations. Stimuli eliciting many social responses may be learned very rapidly in newborn animals by the process of *imprinting*. Imprinting, unlike conditioning and other common forms of learning, is rapid, long lasting, and apparently unrewarded.

Selected Readings

Beer, C. G. 1963–1964. Ethology—the zoologist's approach to behaviour. *Tuatara*, **11**:170–177; **12**:16–39. (A thorough, short, historical review of classical ethology in the form of a critique.)

Klopfer, P. H. 1973. Does behavior evolve? *Annals N.Y. Academy of Science*. (This note deals with the issue of behavior phylogenies and their underlying assumptions.)

Lehrman, D. S. 1953. A critique of Lorenz's theory of instinctive behavior. *Quarterly Review of Biology*. (A paper that touched off controversy and led to a partial synthesis of European ethology with other schools of study.)

Lorenz, K. Z. 1935. Der Kumpan in der Umwelt des Vögels. *Journal für Ornithologie*, **83**: 137–214, 289–413. (Lorenz's most famous paper, in which he introduces the phenomenon of imprinting. A slightly different version of this paper appeared in English: 1937. *Auk*, **54**: 245–273.)

Lorenz, K. Z. 1965. *Evolution and Modification of Behavior*. Univ. of Chicago Press. (This is an expression of Lorenz's more recent views.)

Society for Experimental Biology. 1950. *Physiological Mechanisms in Animal Behavior*. Symposia of the Society, IV. Academic Press, N.Y. 482 pp. (The following chapters in particular typify the viewpoint of the European ethologists: Lorenz, on comparative behavior patterns, pp. 221–268; Tinbergen, on hierarchical organization, pp. 305–312; Baerends, on releasers, pp. 337–360; and Armstrong, on displacement, pp. 361–384.)

Tinbergen, N. 1951. *The Study of Instinct*. Oxford, 288 pp. (The primary source book for European studies and an attempt at a synthesis with the views of other ethologists.)

Part III

CONTRIBUTIONS OF RELATED DISCIPLINES

1900-1970

Chapter 3

LEARNING
AND
INSTINCT
IN ANIMALS

Some Aspects
of Comparative Psychology

Introduction

Man has generally been at the focus of the various studies grouped under the rubric psychology. However, psychologists themselves disagree on their exact goals. To some, psychology is the study of behavior and of the mind; to others, only the former is of interest. For the ethologist, an anthropocentric approach is not essential. However, two areas of psychology are of particular interest: the study of learning phenomena and the laboratory study of animal behavior, often termed *comparative psychology*.

As for studies on learning, many are not directly relevant to ethology, particularly where they deal with tests of theories solely or largely involving man. Even many such experiments using animal subjects are not directly relevant to animal behavior studies, for they often deal with but a single organism, the white rat, which is used simply as a cheap substitute for human subjects (Adams, 1962). On the other hand, Thorpe (1956) has shown that some of this literature does analyze behavior patterns relevant to the experimental animal's normal life and can occasionally be of ethological interest. After all, a white rat running in a maze is similar to a wild rat running in underground tunnels.

Exigencies of space preclude discussion in this book of many theoretic systems of psychology that attempt to explain learning, as well as the animal studies directed principally toward an understanding of man (e.g., McDougall, 1905, 1923; Ebbinghaus, 1913; Guthrie, 1959; Hull, 1943; Spence, 1960). We shall restrict our survey to the major studies of learning which have an immediate relevance to an understanding of animal behavior and to the more interesting of the comparative studies.

Beach (1955) has suggested that problems of comparative psychology ought to originate in the life of the animal studied, the comparison moving "upward" toward man. Heretofore, most work has had the reverse orientation. Nevertheless, some comparative psychological studies of phenomena other than learning are of real relevance to ethology. We have chosen to consider these toward the end of this chapter (also see Chaps. 4, 5, and 7).

Early Experiments on Animal Learning

The thread of interest in animal learning comes from Darwin (and Romanes) through Lloyd Morgan (see Chap. 1) to E. L. Thorndike and

John B. Watson at the turn of the twentieth century. The influence of the latter two workers inspired much of the later study of the learning phenomena.

ANIMALS IN PUZZLE BOXES

If there has been an evolution of the mind, as well as the body, could this not be demonstrated? Thorndike (1911) set about to test this question with a series of problems to be solved by various animals of different species. For instance, he tried to see if chickens could find their way out of a simple maze made of books set on end. But Thorndike relied most heavily on the *puzzle box* to test animals' abilities to solve problems.

A wooden box is so constructed that pushing a lever, for instance, one projecting from the center of its floor, opens the box's door. Since most animals can push such a lever in one way or another, their ability to escape does not require a special motor apparatus; it is simply a problem of the "mind." This method also allows one to test whether animals can imitate one another. An animal that has observed the escape of another from the box can be tested in the same box, and its performance compared with that of the first animal. The difficulty of the problems can be varied by introducing a series of levers that must be pushed in a certain order, or by contriving more complex latches.

What Thorndike found was that animals initially solve this problem pretty much by chance. For instance, a cat investigates its new prison upon introduction into the box; it sniffs, paws at the door, tries to squeeze through the boards on the side, and finally seems to give up. It may then lie down and in the process push against the escape lever. The door opens and the cat rushes to freedom. Upon re-introduction, the cat may very soon lie against the lever again and when replaced in the box repeatedly, come to roll over immediately. This adaptive change in the behavior of the cat upon introduction into the puzzle box is "learning." Cats, however, do not learn how to escape by watching other cats (though they may learn other antics through observation; see review on observational learning by Davis, 1973).

These and other kinds of experiments led Thorndike (1911) to propose his two famous "laws" in his book *Animal Intelligence*:

> *The Law of Effect* is that: Of several responses made to the same situation, those which are accompanied or closely followed by satisfaction to the animal will, other things being equal, be more firmly connected with the situation, so that, when it recurs, they will be more likely to recur; those which are accompanied or closely followed by discomfort to the animal will, other things being equal, have their

connections with that situation weakened, so that, when it recurs, they will be less likely to occur. The greater the satisfaction or discomfort, the greater the strengthening or weakening of the bond.

and,

The Law of Exercise is that: Any response to a situation will, other things being equal, be more strongly connected with the situation in proportion to the number of times it has been connected with that situation and to the average vigor and duration of the connections.

The emphasis upon the "connection" between the situation (stimulus) and the response represents the birth of the concept of *connectionism* in psychology.

Thorndike made two collateral points with respect to animal learning. First, the connections were thought to occur in the central nervous system, specifically at the synaptic connections between neurons. For this he had no proof (see Chap. 4). Second, Thorndike recognized that situations somewhat like the original situation also come to elicit the response. In more recent terms, this phenomenon would be called stimulus generalization (see below).

The difficulty with Thorndike's law of effect lies primarily in the "satisfaction to the animal." Thorndike (1911: 242) explained that "by a satisfying state of affairs is meant one which the animal does nothing to avoid, often doing such things as attain and preserve it." Thus he fell prey to a tautology: Certain responses tend to be repeated because they bring about a satisfying state, the satisfying state being that which the animal repeatedly brings about through its responses.

Thorndike wisely did not interpret the satisfying state or reward as that state that had direct biological advantage. He knew that many animals were satisfied by deleterious conditions (after all, addicting drugs can be very harmful). Sometimes a reward satisfies an evident need, as food satisfies hunger. But, on the whole, Thorndike failed to deliver an explanation of satisfaction that would be acceptable to psychologists.

Thorndike later (e.g., 1932) "repealed" his law of exercise, or at least modified it considerably. The modification stated that repetition of the correct connections did slightly strengthen the bond between stimulus and response, but mere repetition of the entire situation (without regard for errors) did not strengthen the bond. Thorndike also laid increasingly less emphasis on the weakening of bonds through discomfort. Therefore, the essence of his final viewpoint was that reward was the major factor by which bonds are strengthened.

In the work of Thorndike we find nearly all of the major phenomena and problems destined to occupy psychologists for the next half century.

The connections between neurons brought about by learning have still to be discovered. The effective reward that brings about the connection is yet a plaguing problem, some students maintaining that there is always a drive (e.g., hunger) that is reduced by the behavioral result (e.g., swallowing food).

BEHAVIORISTIC APPROACH TO ANIMAL LEARNING

John B. Watson established *behaviorism* in 1913 with the publication of "Psychology as the Behaviorist Views It." His thesis was simple: introspective methods of psychology do not work with animals; one can only infer the nature of the animal mind from the animal's behavior. Therefore, psychology ought to be the study of behavior, not of the mind.

From this beginning came a number of popular books by Watson culminating in the widely read *Behaviorism* (1930). Watson's influence was felt mainly in human, not comparative, psychology when he began to apply "animal" techniques of study to man. He attempted to explain man's behavior as merely a bundle of stimulus-response connections and invoked some simple facts and interpretations of physiology to make his explanations mechanistic. Since Watson spent most of his career outside academic circles and the laboratory, he left few experimental results to illuminate his early ideas. His influence continued to be felt, however, through his brilliant student, Karl Lashley (see Chap. 4).

Learning Paradigms

Psychologists since the turn of the century have attempted to specify and detail the learning phenomena brought to their general attention by Thorndike. Almost all of these psychologists have been behaviorists in the sense that they attempted to reduce the problems of psychology to problems of behavior. Many kinds of learning situations contrived in the laboratory may not represent meaningful situations in terms of the natural lives of the animal subjects, but others are meaningful, as shown by Thorpe (1956). The learning theories of experimental psychology cannot be reviewed here, but certain categories of learning have been studied extensively and are known to apply to many animal species and to operate in their usual lives. These are considered below.

THE CONDITIONAL REFLEX

Pavlov's salivating dog became the topic of household conversation following the English translation of his book *Conditioned Reflexes* (1927).

A Russian physiologist who had won the 1904 Nobel Prize for his work on enzymes, Pavlov had a profound influence on American comparative psychology.

Conditioning Processes

The basic phenomenon discovered by Pavlov is illustrated by his familiar dog experiments. A dog given the taste of food (*Unconditional Stimulus*, or US) salivates (*Unconditional Response*, or UR). Upon repeated occasions a bell is sounded (*Conditional Stimulus*, or CS) just before the food is presented; CS and US are thus paired. After many such paired presentations, the bell alone (CS) can elicit salivation in the absence of food.

The salivation in response to the CS seems identical to salivation in response to the US. However, differences between these responses appear when motor patterns, rather than merely their result (e.g., amount of saliva), are measured. The response to the CS is often of an incomplete nature and is distinguished as a *Conditional Response* (CR). Furthermore, the CR seems anticipatory; the animal may look about as if expecting the US to appear. This anticipatory feature is also brought out by the time of appearance of the CR: if the CS and US are separated by a consistent time interval, the CR appears just before the US is due.

There are special time relationships between the CS and US, as well as between stimuli and responses. In the classic example above, the CS commenced and terminated before the beginning of the US. This is not the only contiguity pattern that will produce conditioning. Konorski (1948) has shown that the CS must end before the termination of the US, otherwise a *conditioned inhibition* to the CS occurs. However, the CS may begin either before or after the US begins and still become effective in eliciting the response (CR) by itself after a number of training trials.

Because the US is necessary to the training, it can be said to "reinforce" the conditional reflex, and thus is a *reinforcement*. We shall see below that in operant conditioning reinforcing stimuli can be dissociated from the unconditional stimuli, at least in one sense. The CS response can become the reinforcing stimulus for a second-order conditioning process. In *secondary conditioning*, a new stimulus, which we shall call CS' (CS prime), is presented with the previously established CS, but without the original US. For example, if CS' were a flash of light given just prior to the bell, the dog would come to salivate in response to the light alone, even though the light had never been associated with food during the training.

How relevant is classical conditioning to observed behavior of wild animals? The question is difficult to answer for a number of reasons. First, the "pure" form of conditioning described is produced only under very precisely controlled conditions. As discussed below, many extraneous factors influence the conditioning process. Because animals live in complex worlds, it is somewhat doubtful whether pure conditioning could take place; more complex, conditioning-like processes may be very common, however. Second, there are only a few studies of conditioning in relation to the behavior of animals in their natural environments (Thorpe, 1956, reviews these). It was once thought by some that chaining of conditional reflexes into second and higher orders would be the mechanism of behavioral ontogeny, but few scientists would ascribe to that view today. One of the reasons they would not is that secondary conditioning is very difficult to produce, even under the controlled conditions of the laboratory.

Some Complexities of Conditioning

A brief consideration of some of the complexities of classical conditioning may demonstrate why *simple* conditioning is difficult to identify in the field. If a novel stimulus (NS) is given at the same time as a well-established CS, the CR does not appear. Pavlov (1927) called this phenomenon *external inhibition*. We have already seen that if the novel stimulus is used in training as a CS', secondary conditioning may occur. That secondary conditioning is difficult to obtain may be partly because of the initial inhibiting effect of the novel stimulus.

Suppose that the novel stimulus is introduced *after* the CS but before the time that the US usually appears. In such a situation, the CR is enhanced (e.g., the dog may deliver more saliva than without the NS). This phenomenon was termed *disinhibition* by Pavlov.

Inhibitory phenomena enter the conditioning process in other ways, too. If a well-established CS is repeatedly given without supplying the reinforcing stimuli (US), the animal soon fails to respond (e.g., produces no saliva to the CS). (In present terms this is called *extinction*; Pavlov called it *internal inhibition* because it is more than just extinguishing the conditional reflex built up previously.) If, some time after the extinguishing process is complete, one again presents the CS, the CR will be shown again, in nearly full intensity. This reappearance of the conditional reflex is called *spontaneous recovery*.

Finally, we may mention *induction*, perhaps the best studied of the more complex influences on the conditional reflex. Suppose that a novel stimulus is given repeated trials in which neither the CS nor the US is given. If the NS is then presented just prior to the CS, an enhanced CR results. Thus *positive induction* is just the opposite of *external inhibition*, for in the latter the NS having not been previously shown in any context inhibits the CR. Furthermore, if the NS is repeatedly shown, as in positive induction, without connection with the CS or US, but introduced in the later test *after* the CS, then no CR appears. This phenomenon is *negative induction*, and it has the reverse effects of disinhibition (where the untrained NS presented after the CS enhances the CR). In studying the acquisition of behavior in the field, the ethologist must be aware of all these complex influences on simple conditional reflexes.

Manipulations of Conditional Stimuli

Two other important phenomena were brought to light by Pavlov: generalization and irradiation. In *stimulus generalization*, stimuli similar to the conditional stimulus become effective in eliciting the response, even though they have not been utilized in the conditioning trials. In Pavlov's words (1927: 113) "for instance, if a tone of 1000 d.v. is established as a conditional stimulus, many other tones spontaneously acquire similar properties, such properties diminishing proportionally to the intervals of these tones from the one of 1000 d.v."

Stimulus generalization in a single individual was demonstrated by Guttman and Kalish (1956) using an operant conditioning technique (see instrumental conditioning below) with pigeons. The bird is trained to peck when shown a given wavelength of light for which it is reinforced with food. Subsequently, no food is offered (extinction period) and various wavelengths are shown in a randomized sequence. The bird's pecking rate during extinction is maximum at the training stimulus wavelength and falls off regularly and symmetrically at increasingly higher and lower wavelengths. Subsequent studies by Guttman, Kalish, and their students have probed the effects of punishment (negative reinforcement) at a second wavelength and other factors. These studies promise to tell us about the processes by which an animal comes to respond in one situation when it actually has been reinforced in quite a different one.

An analogous form of generalization, called *irradiation*, involves the acquisition of conditioning effects by receptors not involved in the conditioning trials. "If tactile stimulation of a definite circumscribed area of skin is made into a conditional stimulus, tactile stimulation of other skin areas will also elicit some conditional reaction, the effect diminishing with increasing distance of these areas from the one for which the conditioned reflex was originally established" (loc. cit.).

The extent to which generalization and irradiation are important in the real lives of animals is unknown. It does seem likely, however, that something like stimulus generalization would be an advantageous capacity if a predator, after obtaining a certain prey in a certain place, would then search for similar prey in similar places.

Pavlov also demonstrated that generalization could lead to uncovering the *stimulus discrimination* capacities of an animal. For instance, to discover how closely a dog can discriminate (or "differentiate," in Pavlov's terms) between similar musical tones, a dog is conditioned to salivate at a particular tone. Then a tone of quite a different pitch is sounded without food-reinforcement. At first the dog salivates a little to the second tone (generalization), but soon ceases to respond at all to it. Then another un-reinforced tone is sounded, this one closer to the conditioning tone. When the dog ceases to respond to this, a third tone yet closer to the conditioning tone is sounded, and so on until a tone is reached that the dog will not cease responding to. The interval between this last tone and the original one represents the discrimination threshold.

INSTRUMENTAL CONDITIONING

In 1938, B. F. Skinner published a work of extraordinary importance for the study of animal learning: *The Behavior of Organisms*. In many ways, the book is an outgrowth of Thorndike's studies. The basic phenom-enon, called *operant conditioning* by Skinner, is but a variant of the old trial-and-error learning by reward such as that exhibited by cats in a puzzle box.

After carefully distinguishing between the Pavlovian conditional reflex and operant conditional behavior, Skinner explores many aspects of the latter using a single apparatus. The apparatus, which has become known as the *Skinner box*, is a living box which contains an object that the animal can manipulate. This object, for the laboratory rat, is a lever capable of being depressed; for a pigeon, a disk (or key) that may be pecked.

The animal possesses a certain, very low probability of responding to the object; that is, pressing the bar or pecking the key. A food hopper may be added and so arranged that its door opens (or a pellet of food is deliv-ered) only when a response is made. Simultaneously the response is re-corded on an electrically activated recorder. Assume a pellet of food is given for each press of the bar. Very rapidly the number of responses per unit increases (*conditioning*) finally leveling off at some maximum. The ap-paratus can be so constructed to deliver a pellet of food only after two or three or any other number of presses of the bar (*fixed-ratio schedule of reinforcement*). Alternatively, food can be delivered only after certain inter-vals of responding (*fixed-interval schedule*).

Note that the term *reinforcement* replaces the traditional word "re-ward" for the food delivered to the animal. This may be taken as symbolic of Skinner's extension of Watsonian behaviorism. That is, Skinner recog-nized that any concept was no more than a complete dossier of the opera-tions used to describe it. This is the "operational" viewpoint of the scientific method which Skinner applied to behavior. According to this point of view, psychology must be concerned with behavior, not mind, since all concepts of mind are reducible to the behavior used to describe the concepts. Since reward and satisfaction connote sensations that might be inferred but could not be verified from measuring the behavior of the rat, the term reinforcement is introduced. Reinforcement is defined operationally: that which alters the probability of the response.

But Skinner's similarity to Watson stops here, for the former rejected the latter's naive "physiologizing." Behavioral processes, Skinner argued, could be studied without reference to the neuronal and hormonal bases that underlie them. This point of view is part of Skinner's operational approach to psychology. Science is primarily a matter of finding out how one variable (e.g., response) is dependent upon another (e.g., stimulus).

An unorthodox aspect of Skinner's book was the insistence that theory has no place in a science of behavior. This outlook is another manifestation of Skinner's operationalism. Since a phenomenon is understood only in terms of operations performed, the psychologist must perform all the possible operations on behavior to obtain a full description. Since all operations will be tried, there is no need to theorize which to try first—just jump in and begin trying. And that Skinner did; the quantity of the experimental results from him and his followers is truly impressive.

Most psychologists do not subscribe to Skinner's atheoretic viewpoint. In a very circumscribed situation (for instance, the Skinner box), such a plan of attack may be suitable, but more normal behavior is just too complex to allow one to perform all the necessary operations. An experimenter's life is finite, and thus only a finite number of experiments can be performed. Some choice between experiments must be made, and the choice will be governed by the expectations of the experimenter. These expectations may, of course, constitute a theory.

The basic phenomenon of instrumental or operant conditioning has been found in a number of representatives of most groups of higher animals (e.g., Thorpe, 1956, who calls it "Type II" conditioning). It is a much more satisfactory unit of learning than the Pavlovian conditional reflex for a number of reasons (cf. Thorpe, 1963a). But, like the conditional reflex, instrumental conditioning has yet to be demonstrated as a widespread phenomenon in the natural lives of animals.

OTHER FORMS OF LEARNING

Most learning-psychologists would eschew, at present, any attempt at a truly inclusive definition of learning. Learning has been studied primarily in a number of given experimental situations, such as the puzzle box, the Pavlovian situation and the Skinner box. There are a number of other learning phenomena that have been elucidated in a circumscribed experimental situation using animal subjects. Brief mention of some of these follows.

Habituation

Habituation is a very simple kind of learning, first distinguished as such by Humphrey (1930) in the German journal *Psychologische Forschung*. *Habituation* is a decrease in the probability of response to a stimulus upon repeated presentation of the stimulus. Humphrey shocked snails repeatedly until they no longer exhibited a withdrawal reaction (drawing in of the tentacles).

Habituation is probably one of the most widespread of learning phenomena (Thorpe, 1956). However, psychologists as a whole have tended to ignore habituation and the phenomenon rarely finds its way into formal theories of learning. We know of its existence in many animal forms primarily through the observations of zoologists. The most common form of habituation appears to involve avoidance (or fear) situations, in which an animal early in its life tends to avoid a very wide range of environmental stimuli, but through repeated exposure ceases to respond to many of these. How long-lasting early habituation really can be is an important question in behavioral development.

Hinde (1960) has further elucidated the nature of habituation by studying the "mobbing" responses of chaffinches to one of their natural enemies, an owl. Hinde finds four separable kinds of *habituation* or response-waning: a rapidly and a slowly recovering motor fatigue specific to the mobbing responses themselves and a rapidly and slowly recovering satiation to the releasing stimulus. The study raises the interesting question as to why chaffinches habituate to a natural predator at all, and it certainly creates new problems for the study of this type of learning. For instance, it provides a tie between learning phenomena and problems of drive (cf. Chap. 2).

Maze Learning

As noted above, mazes were used by Thorndike as a simple test of animal intelligence. A *maze* may be as simple as a "T," where the animal is inserted at the base and must find its way to one of the arms to receive its reward (which may be food or simply freedom from confinement). Upon reintroductions into the maze, the animal repeats its performance increasingly more rapidly, making fewer wrong turns. Mazes may, of course, be made exceedingly complex. The favorite subject of maze studies is a hungry laboratory rat, which takes readily to such problems for food reward.

One result of maze studies has been their demonstration of *latent learn-*

ing. Blodgett (1929) was one of the many early psychologists to show that the rat will subsequently learn more rapidly to run a certain pattern in a maze if previously given opportunity to explore the maze—more rapidly, that is, than a similar rat not given the opportunity for preliminary (unrewarded) exploration. Thus, without patent reward, the rat learns something about the maze in general that facilitates its later learning of a specific path in the maze.

The importance of latent maze learning resides in the difficulties it raises with respect to reward theories. The *drive reductionists* are forced to postulate an "exploratory" drive that is reduced by preliminary exploration. But no simple stimulus-response formulation of learning explains how what is learned in this situation facilitates later learning with food as the reward.

Maze learning abilities are widespread in animal forms (Thorpe, 1956). Latent maze learning may well be the basis of homing behavior discovered in so many animals (see Chap. 8).

Insight

Wolfgang Köhler's book *The Mentality of Apes* was first published in English in 1925. Many ways in which Köhler's captive chimpanzees attained food suggested humanlike mental processes, particularly *insight*. The general concept of insight originated in the largely German school of *gestalt* psychologists; Köhler called insight the *"ah-ha Erlebnis,"* a term which conveys (even in English) the sudden perception of relations that characterizes insight.

Some typical examples of the chimpanzee's insightful behavior given by Köhler are as follows: food, for instance a banana, is suspended from the ceiling by a string; about the cage lie boxes and sticks with which the chimpanzee has had manipulative experience. The chimp may knock the food down with a stick or climb the stick after setting it on end under the banana. (In the latter case, of course, the chimp must climb rather rapidly to be successful.) Some chimps, when the food was too high to reach with a stick, piled boxes under the food and then climbed the pile. In another case, food was set out of reach outside the cage. The chimp was supplied with a stick too short to reach the food, but long enough to reach a longer stick. The animal pulled in the long stick with the short one and then pulled in the food with the longer stick. There are various objections to the conditions of the experiments described by Köhler and their interpretations. However, taken as a whole, Köhler's observations do demonstrate a rather advanced sort of apprehension of spatial relations in the chimpanzee not demonstrable in "lower" animals.

It would seem that insight is only separable from trial-and-error solutions to a problem by the fact that not all the possible (that is, simpler and inadequate) solutions are overtly attempted. It is as if the chimpanzee tries a number of possible solutions mentally until hitting upon a likely one, which is then actually attempted behaviorally.

Insight *sensu stricto* may be considered only that part of the mental process which is concerned with the "apprehension of relations" (Thorpe, 1956), whereas insight-learning is the behavioral solution of the problem.

Learning Sets

Harry Harlow's (1949) concept of the *learning set* relates simpler trial-and-error learning to "higher," insightful-like, learning. A monkey is shown a tray with two differently shaped boxes, one of which has food hidden beneath it. The monkey lifts or turns over one of the boxes; the response is either correct (and the monkey gets the food) or incorrect. In a typical experiment the monkey is given five more trials with the same pair of boxes, the spatial positions of which are randomized; the same-shaped box always contains the reward. When the results of a number of naive monkeys are averaged, a steady improvement in the percent of correct choices is seen to occur over the six successive trials (at the first trial the mean score of correct responses is, of course, 50%). This is a standard discrimination-learning paradigm in which the increase in percent correct responses is linear or sigmoid with the number of trials.

The learning set phenomenon is shown by presenting a new set of two boxes to the same monkey. This new set is also given for six trials, but discrimination-learning proceeds more rapidly than with the original pair of boxes. For the following set of six trials, a third set of boxes is used, and so on. By the time a hundred or so trials have been given, it is clear that the discrimination-learning curve is no longer sigmoid or linear. Beginning, of course, at 50% correct on the first trial of any of these later series, the mean percent of correct responses on the second trial is phenomenally high (80–95%). The increase between trials two and six in these later series is gradual (Fig. 3–1). In other words, the choice goes from random to nearly perfect discrimination as the result of one trial; but the antecedent experience necessary for this learning capability is a group of very similar discrimination problems.

The learning set phenomenon has been termed *learning-to-learn*, although it is probably more parsimoniously thought of as a transfer of discrimination of all factors in common in all the trials (that is, everything except the shapes of the boxes). The learning set phenomenon is particularly important because it demonstrates how the accumulation of experience during the lifetime of an animal facilitates its ability to "learn" in a "new" situation since few situations will be truly new in all respects.

Reinforcement

Turning from learning phenomena to the analysis of reinforcements that mediate them, several approaches are popular. Olds (1958), on the one hand, and Brady and Nauta (1955), on the other, have found *reinforcing centers* in the brain. A chronically implanted electrode can deliver a shock to the brain when a lever is depressed by the animal. Some electrode placements bring about a tremendously high continuous rate of self-stimulation by rats and other animals while others bring about no responsiveness at all—in fact, the animals may exhibit "fear" responses after pressing the lever.

Using a highly controlled artificial environment, Kavanau (1963) has shown that animals will work and learn for mere control over some part of their environment. Finally, a more probable behavior pattern will reinforce a less probable one (Premack, 1963). For instance, if a rat must drink a certain amount of water in order to free a running wheel, its rate of drinking will become very high—as in the ordinary instrumental conditioning situation. If, on the other hand, access to water is made contingent upon running, the animal will run very much more than usual in order to drink. The phenomena now being dealt with by Kavanau and Premack have yet to be demonstrated in the natural lives of animals.

CONCLUDING REMARKS ON LEARNING THEORY

An entire book could be devoted to a more detailed exposition of psychological learning theory—in fact, many books have been devoted to this subject. Most of the major problems that now confront learning theorists were raised in some way by Thorndike and most have been attacked with operational methods that stem from Watson. Learning phenomena studied in most detail by psychologists are isolated laboratory paradigms considered, but not yet proven, to be of widespread importance in the natural lives of animals. Many other less well-studied phenomena are reported in the psychological literature, and a number of important theoretical formulations of learning in general have been attempted (e.g., Tolman, 1932; Guthrie, 1959; Hull, 1943; Spence, 1960; and Miller, 1959). The neural basis of learning is considered in Chap. 4.

Other Problems of Comparative Psychology

The problems other than learning theory with which comparative psychologists have dealt are closely related to the study of learning, for instance, the phylogeny and ontogeny of learning abilities. Other psycholog-

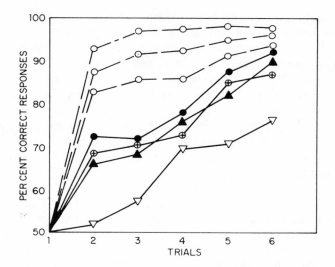

Fig. 3–1. The four lower curves illustrate the shape of learning curves when the problem (a 6-trial nonspatial discrimination) is first presented. The three upper curves, which are discontinuous and have two distinct components, reflect learning rates after the subject has had considerable experience with similar problems. (After Harlow, 1949.)

ical problems, such as the physiological mechanisms of animal behavior, require separate discussion (see Chaps. 4 and 5).

THE PHYLOGENY OF BEHAVIOR AND LEARNING

The problem of tracing the evolution of the mind that spurred Thorn-dike's experiments has continued to be of some concern to comparative psychologists. Maier and Schneirla (1935) attempted to provide a phylogenetic basis of mammalian behavior in the first part of their book *Principles of Animal Psychology*. For each of the major animal groups they provided a synopsis of the neural and muscular mechanisms of behavior and discussed how structure was correlated with the known behavior of the animals. Much of their information on lower animals actually came from the works of Jennings or Romanes (see Chap. 1) on the one hand or from natural history observations by zoologists on the other, but some came from psychological experiments as well. The integration of these sources provided an important impetus for psychological experimentation on nonmammalian subjects. However, the fact that this book remains the most important summary of this literature for psychologists indicates the decline of truly comparative psychology since the first quarter of this century.

Increased complexity of behavior and increasing capability for behavioral modifications do not strictly follow modern phylogenies of the animal kingdom. Part of the reason is that some animals high on the scale of phylogeny (i.e., groups evolving late in geological time) live in the sea (e.g., starfish) while others have invaded land (e.g., insects). Insects show more complex kinds of behavior than starfish because they must interact with a more complex environment on land than do echinoderms in the sea. For each group of organisms, Maier and Schneirla list the receptor equipment and the stimuli to which the animals respond, the conduction apparatus that communicates between the receptors and effectors, and the movement and action effector apparatus. Nonmammalian animals are discussed under eleven rubrics.

The first three categories embrace protists (and plants!), the first multicellular animals (sponges and coelenterates), and advanced radially symmetrical animals (starfish and their relatives). Multicellular animals face the problem of co-ordinating many parts of an animal in response to specific environmental influences that impinge simultaneously at specific points on its body. However, coelenterate animals have no special equipment for integration of incoming stimuli so that behavior (e.g., movement) is controlled by simple interactions of the separate receptor-effector systems. Echinoderms have ganglion cells in a nerve ring that facilitate communication between separate arms, although they do not exercise centrally determined control over behavior.

In sum, this series of animals shows an increasing specialization of behavior and of its mechanisms: the development of receptors and sense organs, contractile tissues, nerve nets and tracts, increasing amounts of connecting neural material between the increasingly superficial receptors and increasingly internal effector systems.

The second four categories (two on arthropods) include flatworms and annelid worms, molluscs (snails, clams, squids, etc.) and arthropods; with a separate chapter on orientation, learning, and social organization of insects. Here the same trends noted in radially symmetrical animals continue, with the added properties of a bilateral symmetry and, concomitantly, a bilateral nervous system and the development of cephalization. The segmented, bilateral animal is capable of finer adjustments in behavior; thus the animal can cope with the more complex environments of terrestrial life. Cephalization allows finer integration of sensory information as well as centrally controlled behavior.

Finally, Maier and Schneirla discuss nonmammalian vertebrates in separate chapters on fish, amphibians, reptiles, and birds. Here all the previous trends are seen to continue, including especially the development of a large and complex central nervous system interposed between receptors and effectors. The amphibian's transition to land brings the greatest acceleration in improvements in behavior and its underlying mechanisms—differences observed even between the aquatic and terrestrial stages of the same species.

Taken as a whole, the development of more complex behavior and the development of abilities to alter behavior adaptively through experience are correlated with an increased specialization of separate parts into sense organs, integration centers, and so on. Although much of the information, both structural and behavioral, presented by Maier and Schneirla could now be augmented, the general trends they outline are still valid.

That "phylogeny of learning" should not be considered a unitary evolutionary process has been emphasized by comparative psychologists such as Schneirla (e.g., 1959). His studies of ants, for instance, lead to the conclusion that maze-learning in ants and rats really represents different kinds of processes. An ant builds its maze habits in serial sections which show no integration as a whole, whereas the rat anticipates a distant goal and can get to it more rapidly if a shorter path is opened for its use. After learning one maze, a rat learns another one more quickly (see Learning sets, above), whereas an ant's performance in a new maze is hampered by its previous experience with the first one.

Schneirla (e.g., 1949) views the increase in learning ability that correlates with certain structural changes in animal evolution to be the adding of new kinds of learning rather than a quantitative increase in general learning ability. Protozoans are capable of modifying only sensory parameters such as threshold as the result of experience. Flatworms show habituation, and other simple but centrally mediated changes, in response to stimulation as the result of experience. Segmented worms, however, can show true association conditioning (or contiguity-type conditioning in Schneirla's terminology). More subtle forms of discrimination learning and trial-and-error learning appear in terrestrial animals living in a complex environment (insects, birds, mammals), whereas insight and other higher processes are restricted to higher mammals.

Schneirla's views are consonant with classical ethological studies (cf. Chap. 2) in that he too assumes that the animal is studied with regard to the environment in which it evolved and for which its behavioral capacities were selected to serve.

LEARNING AND THE ONTOGENY OF BEHAVIOR

Although Thorndike and other early experimental psychologists were aware of "instinctive" behavior patterns in animals, the notion that nearly all behavior was learned arose from the writings of Watson and other psychologists who were preoccupied with learning phenomena. Surprisingly little experimental work on the ontogeny of behavior in animals was undertaken by psychologists. A brief look at some notable exceptions follows.

Maier and Schneirla (1935) make a number of important points about the ontogeny of behavior. For instance, the appearance of a response at birth says little about its development, first, because some animals have longer gestation or incubation periods than others and may thus be expected to show greater complexity of behavior at birth or hatching, and second, because experience of some kinds may be effective in structuring the behavior of the embryo. They also make the point relevant to classical ethology (Chap. 2) that species-specificity of behavior can be due to experiences by all individuals in similar environments and thus innateness cannot be logically inferred. The subtle formulation of Maier and Schneirla that all behavior is the product of heredity and environment or their interaction has been sometimes overlooked in ontogenetic studies.

Domestic chicks peck very soon after hatching from the egg. Thorndike, Spalding, and other observers at the turn of the century published experiments and observations demonstrating the remarkable perfection of this behavior. Shepard, Breed, Bird, Padilla, Cruze, and other psychologists tried to determine if pecking could be very rapidly learned (see Cruze, 1935, for a list and criticisms of the earlier experiments). Dark-reared chicks (in some cases force-fed, in some cases unfed) were placed on a surface with grain scattered about and the accuracy or completeness of their pecking noted. Although the various experiments are in some measure contradictory, one generalization about the ontogeny of pecking can be made: some development is due to experience and some is not. For instance, in chicks dark-reared for five days and then given ordinary experience with grain, the percent of swallowing reactions rises to the same level (about 80–90%) as in chicks reared in light from birth. But the dark-reared chicks learn the swallowing reaction much more rapidly. However, if force-fed in the dark too long, chicks develop abnormal feeding movements and never develop normal pecking behavior.

Could reflexogenous movements "teach" the organism more complex responses? Zing Yang Kuo (e.g., 1932) removed parts of the shells of eggs of domestic chickens and watched the movements of the embryos by smearing Vaseline on the membranes to make them transparent. He reported that the beating of the heart forcibly moves the chick's head, which lies over the chest. The movement in turn activates the neck muscles, which contract with increasing autonomy. These sorts of embryonic movements, Kuo argued, were precursors of post-embryonic pecking movements and served to "train" the animal. Kuo's speculations were not tested experimentally for many years, though there were historical precedents. In 1885, Preyer authored a study of the embryogenesis of movements of the chick. From the later 1920's onward there appeared occasional studies on the spontaneity and development of movements, particularly by Coghill (1929) and Carmichael (1934). Much of this work has been reviewed by Gottlieb (1970). He concludes that the evidence supports the notion of spontaneous and sustained (i.e., independent of sensory input) embryonic motility. This in itself hardly suffices to either support or refute Kuo's form of behavioral development. But, as Hamburger (1968), one of the principal workers in this area, points out:

> One might have wished . . . that behavior in all vertebrates is integrated from beginning to end. Instead, we are confronted with a puzzling diversity of phenomena that are difficult to fit in a coherent theory. . . . The idea of autonomous neural differentiation that proceeds according to an intrinsically determined program has won over the rival idea that adaptive neural connections are the result of selection, by trial and error, from a randomly interconnected network; the role of spontaneous motility . . . has been recognized . . . [but] . . . on the behavioral level, the speculations about storage of prenatal 'experiences' and their influence on postnatal behavior are, up to now, without critical experimental foundations. (P. 267.)

Finally, Oppenheim (1970), who actually removed segments of the limbs of duck embryos, specifically tested Kuo's hypothesis on the co-ordinating role of reflex responses and found it wanting.

Kuo also concluded that cats must learn to kill rats, i.e., that what is regarded as species-characteristic ("instinctive") behavior is in fact learned. Some of his results were (1930: 30):

> Of the 21 kittens raised in the rat-killing environment 18, or more than 85 percent, killed one or more kind of rats before four months old. The kittens killed the kind of rat which they saw their mothers kill though they might kill other kinds of rats as well.
> Of the 20 kittens raised in isolation only 9 (or 45 percent) killed rats without the so-called learning.
> All kittens raised in the same cages with rats never killed their cage-mates, though 3 out of 18 killed other kinds of rats.
> Of 11 non-rat-killing kittens 9 became rat-killers after seeing other cats in the act of killing rats. But with the exception of one kitten the re-enforcing stimulus of seeing other cats killing rats had failed to make the kittens raised in the same cage with rats follow the same action.

What Kuo's results show, of course, is that experience can facilitate rat killing, but his study does not answer the question "Why do cats reared in isolation kill rats?" The response of the classical, European ethologists would obviously be in terms of an IRM. However, apparently even this extreme form of preformationism may yield to some of Kuo's suggestions. At an international conference of ethologists in 1964, Lorenz proposed an "innate schoolmarm" as a solution to the question of how stereotyped, species-characteristic behavior developed. Though intended facetiously, Lorenz' example is of interest: imagine a larval fish, curled within the egg membrane, its tail opposite its eye. The spontaneous twitching of its tail (*pace* Kuo and Hamburger!) assures the development of a subsequent orientation vis à vis conspecifics that leads to a characteristic school, and associated schooling behavior. This notion is interesting because it can be tested. More on this subject will be said under Some Recent Developments.

Psychological experiments on the ontogeny of visual discrimination have also been performed on a number of animal species (e.g., Lashley and Russel, 1934; Riesen, 1947) by dark-rearing animals from birth. However, it is possible that in experiments such as these some retinal degeneration takes place if the eye is not stimulated by light (Weiskrantz, 1958). Thus any subsequent lack of abilities to discriminate visually is difficult to interpret. Lashley's study on the rat, however, did show that depth perception and light-dark discrimination were normal in dark-reared animals; Oppenheim's (1968) work on the role of visual stimulation of embryos on later color preferences and Gottlieb's related studies on sound preferences lead to somewhat different conclusions (see the recent developments discussed below).

In conclusion, it seems clear that the development of animal behavior cannot be explained merely in terms of the usual learning processes enumerated at the outset of this chapter. Pavlov required for classical conditioning the unconditional stimulus and unconditional response, and Skinner required for instrumental conditioning an operant response with at least a low probability of being directed to the manipulative part of the apparatus. Likewise, all the other formulations of learning theory have had to begin with some kind of behavior which is altered through the learning process. It would seem, then, that the study of ontogeny should be directed not to the question "is behavior instinctive or learned?" but to the question "how does behavior develop?"

BEHAVIORAL INHERITANCE

Since the nervous and other structural systems that determine behavior develop from the fertilized egg as do all systems of the body, it is natural to inquire how breeding manipulations affect behavior. It is expected that actual behavior (a phenotypic expression of nervous structure) may not bear as close a correspondence with the genetic endowment of the animal as do other phenotypic characters because behavior can be so greatly influenced by the environment during development. Nevertheless, using the geneticist's method of holding the environment constant while varying the genetic constitution through breeding, it is possible to study the inheritance of behavior.

Some general conclusions reached toward the end of our survey period are reviewed by Hall (1951). To begin with, one can readily show differences in behavior between rodents of different genetic strains reared under similar conditions; for instance, some are generally wilder and less tractable than others. Strain differences measured in aggressiveness, temperature preferences, and susceptibility to convulsions induced by sounds have been studied by Scott, Hall, and others. In the mouse, *Peromyscus maniculatus artemisiae*, it appears that susceptibility to audiogenic seizure is inherited as a single recessive gene; but most behavior patterns or learning abilities studied are the product of multiple genes whose effects are more or less additive.

An early critical study of behavior genetics was carried out by Tryon (1940) and his co-workers. They claimed to have shown that maze learning in the rat is an inherited ability. McDougall (1938) had proposed that maze-learning was an example of Lamarckian inheritance of acquired characters since some of his rats trained in a maze gave birth to offspring whose performance was better than the offspring of untrained rats. A genetic study extending over many years proved McDougall incorrect, although the study raised some problems yet to be fully explained. In repeating McDougall's experiments, Tryon found that sometimes offspring of trained rats did actually show better maze performance than controls, but sometimes they showed worse performance. In other words, over successive generations, the genetically determined maze-learning ability *drifted* toward better or poorer ability; however, the drifts were not correlated with the experience of the parental generation and may have been the product of subtle selective pressures or of simple statistical variations in gene recombinations.

The interaction of heritable and environmental factors is illustrated by a series of studies of the behavior of mice reared with rats (Denenberg et al., 1964) and a study of habitat preferences by Wecker (1963). In the latter study, prairie-loving and forest races of a single species of deer mouse (*Peromyscus sp.*) were trapped in their respective habitats and then tested for their relative preference of a simulated "woods" and "grassland" environment. Lab-reared (twelfth-generation) offspring of one-time field mice were also tested. If we consider the prairie race, offspring of the recently trapped group, they regularly selected the field over the woods irrespective of whether they were reared in a simulated "woods" or "field." Woods rearing weakened the field preference, however. The twelfth generation of lab-reared mice showed a preference for the "field" only when they themselves had been reared therein; for these animals the genotype no longer specified the preference.

In conclusion, it seems likely that genetic components can be demonstrated through breeding experiments to underlie many behavioral patterns and abilities, but that these genetic components will rarely appear as simple Mendelian ratios. This is because the development of behavior is a mosaic composed of so many factors that single detrimental gene mutations are swamped by many other genetic components and overridden by environmental influences.

PERCEPTION IN ANIMAL PSYCHOLOGY

One interesting problem with which comparative psychologists have dealt is uncovering sensory capacities of animals by means of conditioning. For instance, Lashley (1916) studied color discrimination in domestic chicks and Hamilton and Coleman (1933) did a careful study of color discrimination in the pigeon. A series of such studies by psychologists revealed that birds and fish have quite good color vision, but that mammals (excluding primates) possess poor if any color-discrimination capacities.

Discrimination was extended to other forms of sensory input as well. Lashley invented a *jumping stand* in which rats had to leap across a chasm to one of two doors. The correct door with a certain visual pattern (e.g., triangle) falls back onto a ledge when pushed, but the wrong one with a different pattern (e.g., square) is locked so that the rat falls to the floor. Studies such as this showed that pattern discrimination is well-developed in both birds and mammals, but that patterns may be evaluated differently by different species. This general method of testing sensory capacity has also been developed in operant conditioning, and sensory capacities have also been tested independently by zoologists (e.g., von Frisch; Chap. 6). Many sensory capacities have been tested for discriminability: brightness, tone, size, and so on. An important concept of the *equivalence of stimuli* was developed by Klüver (1933). If rats are trained to choose, for instance, a large square in preference to a small square, they will choose in a subsequent test the small square in preference to a yet smaller one. In other words, they learn the difference or size relation between the original stimuli, rather than the size *per se*. The second set of squares are called equivalent stimuli.

Pavlov (1927) found that dogs would develop *experimental neuroses* when trained in a certain discrimination through classical conditioning. If the discrimination is between a circle and an ellipse, for instance, and subsequent to training the ellipse is made increasingly like a circle, the discrimination between the two breaks down. At this point dogs become either intractable or completely unresponsive and lethargic.

Finally, the studies of octopus learning deserve mention because these subjects cannot learn certain kinds of discriminations, for instance, a left-tilting versus a right-tilting oblique rectangle. Sutherland (1959) has tried to explain the discriminatory abilities and limitations of the octopus in terms of neural models of the retina that are consistent with present physiological evidence, although no specific model created to date has predicted the experimental results completely.

These brief examples show some of the capabilities that animals exhibit when tested experimentally by means of learning processes. Certainly learned discriminations become a part of many animals' normal behavioral repertoires, but the extent to which these phenomena are important in normal behavior has yet to be elucidated.

Some Recent Developments

The concepts of learning we have entertained for the past decades are certain to be overthrown or, at the least, radically modified, if the most recent developments are a portent. Four particular, and interrelated, lines of inquiry are taking shape. The most circumscribed of these is a re-awakening of an interest in observational learning and imitation, with particular attention to the social roles and status of individuals and their effectiveness as "teachers," for example, Chesler's (1969) demonstration that kittens learn more from their own mothers than from other cats. Much of this work, and its status, has been reviewed by Davis (1973). This line of inquiry connects studies of dominance and social status (note review by Rowell, 1973, which questions the unitary basis of "dominance") with more conventional studies of learning and reinforcement. One important inference from studies of observational learning is that processes akin to self-recognition and empathy occur during much learning, clearly implying a deficiency in simplistic S–R notions.

The second, converging, line of inquiry relates to internal constraints on learning. This last phrase implies the existence of internal restrictions on the effects of the external environment. These restrictions are diverse and are not limited to sensory or perceptual filters. They also involve processes which search for and control stimuli emanating from the environment. Newly hatched ducklings will follow a wide range of objects that are presented to and moved before them. The longer their exposure to a particular object, the more they will ignore or flee from different objects—for a time.

Subsequently, however, novel or dissimilar objects may elicit an approach in preference to the original, familiar one. "As the birds learn more about the familiar, objects which are detected by them as being slightly novel will in reality resemble more and more closely the familiar object" (Bateson, 1973, in press). This class of phenomena also requires an explanation with other variables than those permitted by conventional learning theories. (Similar problems are considered in the volume by Hinde and Hinde, 1973.)

A third line of inquiry actually includes, or wholly converges with, the foregoing. It is directed to the issue of the "context" in which learning occurs. Context implies a number of boundaries. For instance, the ecology and history of a species are one set of boundaries. If one ignores the natural history of lab rats, it is hardly possible to make sense of the fact that in a maze-reversal task (where the animal must learn to alternate the direction it chooses in order to reach the goal), food is an effective reward but water is not. If one considers the normal environment of the lab rat's forebears, one recognizes that food generally has a variable location, but water is more often fixed to a particular site. Natural selection could then be expected to lead to an internal constraint on learning which made food and water differentially effective as reinforcers. Thus, the past natural history provides a context that cannot be ignored in understanding learning by rats (Garcia, 1973; Rozin and Kalat, 1971).

Finally, the stream into which all four currents merge represents a reassessment of the role and status of the field of comparative psychology. Is a unified view of animal behavior possible or are we forever to be left with "a confused scatter of views of nature, problems, and methods" (Lockard, 1971)? Darwin had argued for the continuity of mental development; but recently, as implied above, the importance of convergence has become emphasized. "Since each animal species evolved unique and separate adaptations to its own niche, and each niche is unique, it would at first seem that a science of behavior could not succeed beyond catalogs describing all the independent behaviors of all species. Generalizations about seemingly valid categories of behavior such as parental care or hoarding or social organization are unlikely to prove fruitful because each kind of behavior evolved independently a number of times . . ." (Lockard, ibid., p. 172). "The old concept of an animal as having some degree of intelligence and thus able to learn nearly anything in accord with its endowment is giving way to the view that natural selection has probably produced rather specific learning mechanisms that correspond to ecological demands . . . (ibid., p. 174).

Despite the pessimistic outlook for valid and yet broad generalizations about learning and behavior that might be expected, it appears that a new level of analysis is being brought to bear, one that extracts general principles even while acknowledging the context-dependency of learning.

Overview

Thorndike found that animals escaped from *puzzle boxes* pretty much by trial and error but escaped more rapidly and with fewer errors upon reintroduction into the same box. His *Law of Effect* said that rewards promoted this sort of improvement in escape and that punishments inhibited them. His *Law of Exercise* said that mere repetition will also bring about such learning. Watson created *behaviorism* as the operational study of behavior without the intervening variables of "mind."

Subsequent experimental psychologists studied more closely operational aspects of learning, with special reference to rewards. The *conditional response* of Pavlov is the linking of an originally indifferent stimulus (the CS or *conditional stimulus*) to a slightly modified *unconditional response*. This is accomplished by presenting the CS co-temporally with the unconditional stimulus that usually elicits the response—and rewarding the animal immediately thereafter. Skinner found that "any" stimulus could be linked to "any" response by *reinforcing* the animal with a reward such as food whenever it responded during the stimulus presentation.

Other forms of learning include *habituation*, the waning of a response upon repeated presentation of the eliciting stimulus, and *latent learning*, in one example of which the animal subsequently learns the path to food in a maze more rapidly and with fewer errors if given prior exploratory experience in the maze without food or patent rewards. Köhler described many examples of sudden and improbable problem solutions by chimpanzees, phenomena called *insight-learning*. Finally, Harlow found that animals given many trials become sufficiently familiar with a given experimental situation that they can learn in only one trial subsequent discrimination problems, this being the phenomenon of *learning set*.

Comparative aspects of behavior show that learning abilities in animals are correlated with increasingly complex nervous systems, particularly with increasing neural material between receptors and effectors, bilateral symmetry, segmentation, and cephalization. Ontogenetic studies of natural behavioral development are few, but they tend to show that the animal is not a *tabula rasa*, but that usual kinds of learning phenomena are intertwined with behavioral structuring processes independent of specific experience.

Selected Readings

Hinde, R. A. 1970. *Animal Behavior*. McGraw-Hill, New York. (This is a difficult, technical, but certainly the most comprehensive and synthetic treatment of the field, and especially of the issues of drive and motivation.)

Koch, S. (ed.). 1959. *Psychology: A Study of a Science*, Vol. 2. General Systematic Formulations, Learning, and Special Processes. McGraw-Hill, New York. [The volume includes separate reviews of the principal learning theories: see especially the chapters by Tolman, Guthrie, Miller, Logan (on Hull and Spence), Skinner, and Harlow.]

Maier, N. R. F., and T. C. Schneirla. 1935. *Principles of Animal Psychology*. McGraw-Hill, New York. (Their review of the correlation between structural complexity and behavioral complexity has been brought up to date in a simpler form by V. A. Dethier and E. Stellar in their *Animal Behavior: Its Evolutionary and Neurological Basis*, published in 1961 by Prentice-Hall, Englewood Cliffs, N.J.)

Thorpe, W. H. 1956. *Learning and Instinct in Animals*. Harvard University Press, Cambridge, Mass. (This is by far the most important historical account of learning for ethologists because it includes not only a review of major learning paradigms but also a survey of learning studies related to natural behavior in all the major animal groups. A shorter account was given by Thorpe in the fourth *Symposium of the Society for Experimental Biology*.)

Chapter 4

THE

ORGANIZATION

OF

BEHAVIOR

The Central Nervous System and Behavior

Introduction

It was not always so, but by the nineteenth-century the notion that the brain is the locus of mental activities was acceptable to most biologists. Some of the interesting background of this subject (see Boring, 1957) really antedates our survey period, but should be mentioned nevertheless. The nineteenth-century phrenologists, under the leadership of Franz Joseph Gall, had earlier given a crude empirical support to this *materialistic view* through their analyses of character and ability. Phrenologists inferred a man's character and ability from the superficial structure (bumps) of his skull (and thereby, presumably, the underlying brain). This inference depended on an association between brain structure and personality (Fig. 4–1). At least some of the phrenologists' clients felt such analyses to have been sufficiently reassuring so that it can be claimed for phrenology that it has provided public support for a materialistic view of mental phenomena.

Direct experimental evidence of the role of the brain in mental processes was apparently first provided by Pierre Flourens in 1824–1825, who, as the consequence of careful extirpation studies, was able to assign functions to various divisions of the central nervous system: cerebellum, medulla, and corpora quadrigemina. He also emphasized the fact that these regions, far from functioning independently of one another, interact in complex fashion. To describe this, he coined the phrase *action propre et action commune*, expressive of the principle that while each portion of the brain may have a specific function delegated to it, it also participates in other functions that represent integrated activities of the whole brain. The resemblance of this notion to certain of Lashley's views, expressed a century later (and which are considered below), cannot fail to impress one.

In the latter portion of the nineteenth-century, students of the nervous system concentrated in four major analytical approaches: (1) the gross anatomical and histological structure of the central nervous system, including changes as consequence of age and experience, (2) the nature of nervous activity and information storage and control, (3) the effects of brain damage on behavior, particularly learned behavior, and (4) the study of localized chemical events that could be correlated with overt behavioral or electrical changes. All these approaches were ultimately directed to localizing particular functions within specified components or processes of the nervous system. After considering these approaches and their results up to 1970, we shall deal with a few more specific problems of brain function.

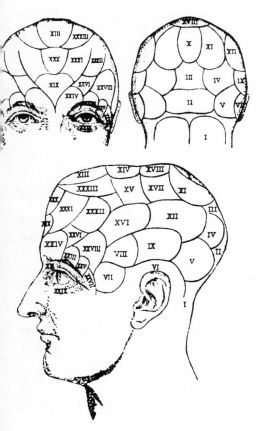

THE

PHYSIOGNOMICAL SYSTEM

OF

DRS. GALL AND SPURZHEIM;

FOUNDED ON

AN ANATOMICAL AND PHYSIOLOGICAL EXAMINATION

OF THE

NERVOUS SYSTEM IN GENERAL,

AND OF THE

BRAIN IN PARTICULAR;

AND INDICATING THE

DISPOSITIONS AND MANIFESTATIONS OF THE MIND.

By J. G. SPURZHEIM, M. D.

Being at the same Time a Book of Reference for Dr. Spurzheim's Demonstrative Lectures.

ILLUSTRATED WITH NINETEEN COPPER-PLATES.

THE SECOND EDITION GREATLY IMPROVED.

LONDON:

PRINTED FOR BALDWIN, CRADOCK, AND JOY,

47,. *Paternoster Row;*

WILLIAM BLACKWOOD, EDINBURGH; AND J. CUMMING, DUBLIN.

1815.

Fig. 4–1. Title page and frontispiece of an early volume on phrenology. In attempting to correlate mental "faculties" with swellings of particular regions of the head, phrenologists drew diagrams of this type; a debased version of this type of activity now occurs at fairs. The claims of the phrenologists were not justified by their data, but they did draw attention to the need for a scientific study of criminology. Reaction against the far-fetched claims of the phrenologists led to the view, in the second half of the nineteenth century, that the brain "acts as a whole." (From Walsh, 1964.)

Neurology

The impact of evolutionary questions on anatomy was manifest in the aim of comparative anatomy: to elucidate the phylogeny of animals and animal structures. This aim naturally extended to the brain itself. Since certain brain structures were more highly differentiated in some animals than in others, structural differences were correlated with the differences in the animals' general behavior.

GROSS ANATOMY

The gross anatomy of the vertebrate brain is best seen during embryogeny, when the anterior part of the spinal cord becomes large. A ventral trough appears in the swelled "brain," dividing the anterior *forebrain* (prosencephalon) from posterior parts. A small ventral trough divides the latter parts into an anterior *midbrain* (mesencephalon) and a posterior *hindbrain* (rhombencephalon). These are the three major divisions of all vertebrate brains.

The forebrain then differentiates into an anterior and posterior part (telencephalon and diencephalon, respectively), as does the hindbrain (metencephalon and myelencephalon, respectively). Thus, there come to be five important areas of the brain. The principal structures in these five areas (from anterior to posterior) are: (1) the olfactory bulb and paired cerebral hemispheres, (2) the thalamus and pituitary (part of which is not embryologically brain tissue), (3) the tectum, (4) the cerebellium (and part of the medulla), and (5) the medulla. In various vertebrates these parts become prominent or decrease in size, and considerable folding of some parts over others occurs.

In lower vertebrates, olfactory sensory information goes to the forebrain, visual information to the tectum, and auditory (and related) information to the hindbrain. In mammals, visual information is instead diverted from the midbrain to the cerebrum.

Gross behavioral correlations with these gross areas can be made. Birds, which depend on quick reflexes in flying, have relatively large cerebella. Mammals having an acute olfactory sense also have generally well-developed olfactory lobes, while visually oriented frogs have large optic lobes, etc. Man, of course, has gigantic cerebral hemispheres, correlated with his superior learning abilities, so that it is the hemispheres that receive much attention from neuropsychologists (see below). Since this gross level of analysis cannot yield very specific facts about the central control of behavior, however, it is necessary to turn to histological studies.

HISTOLOGICAL STUDIES

Histological studies (details of which have been summarized by Boring, 1957) soon led to a recognition of the high degree of differentiation to be found within the mammalian cerebrum and brain in general. These studies encouraged a search for discrete functions which could be assigned to the distinguishable layers or areas of the brain. In 1861, Paul Broca announced the discovery of a center specific for the control of speech. An intelligent though inarticulate patient he had examined proved to have a localized tumor in what has since been known as *Broca's center*. The subsequent discovery by G. Fritsch and E. Hitzig that limb movements in both dogs and men could be elicited through electrical stimulation of the cortex gave impetus to the trend Broca helped to initiate. Systematically exploring the exposed motor cortex of dogs and rabbits, Fritsch and Hitzig found that particular movements were associated with the stimulation of specific regions of the cortex. Some years later, D. Ferrier (1876) and H. Munk (1890, see discussion in Boring, 1957) discovered the existence of specific sensory centers, a discovery that appeared to enhance the view that the brain was to be thought of as an organ compounded of discrete centers, each with a specific (and at least partially autonomous) function.

In an earlier decade (1838), Johannes Müller (and, independently, Sir Charles Bell) had proposed the doctrine of specific nerve energies. Five kinds of nerves were believed to exist, each differing qualitatively from the others in the nature of its impulse. H. L. F. Helmholtz (1852), following the earlier lead of T. Young, extended this notion to qualities within specific sensory modalities. Thus, in vision, the stimuli from different wavelengths of light were thought to be mediated by qualitatively different photo-receptors. This accorded well with the notion of the brain as a histological and functional mosaic.

The interest in and emphasis upon histologic structure, combined with central localization, and the qualities of the peripheral sensory input alluded to above, yielded an intellectually satisfying picture of brain function. It was a picture of the brain as a mosaic of discrete centers, each supplied by nerves whose impulses differed qualitatively according to the character of the stimulus and the nerve. That it was an inadequate picture was only revealed gradually as experiments, such as those by Sheperd I. Franz and Karl S. Lashley in the first two decades of the twentieth-century, demonstrated the inexact and temporally variable nature of cerebral localization (below). The nature of the nervous impulse (below) revealed inadequacies in the doctrine of specific nerve energies. Also left unanswered was the basic question: If the brain does consist of discrete centers, what is a center? Is it a Cartesian *sensorium* or merely a bottleneck through which certain impulses or excitation must pass (Boring, 1957)?

While purely histological investigations of nervous tissue continue, the major results of these investigations have already been reviewed by Ariëns Kappers et al. in 1936. The contemporary trend is to correlate neurology with specific functions, either electrical activity of the brain or overt behavior after brain destruction. The great resolving power of the electron microscope (EM) has permitted drawing correlations between changes in behavior and changes in the fine structure of neural tissues. In many of the studies discussed below, EM techniques were used to determine precise tissue changes which accompany, for example, neural injury and maturation.

Electrical Activity of Neural Structures

Investigation of electrical activity of the brain has become technically sophisticated only rather recently and thus few illuminating conclusions can be drawn from studies prior to 1950. Three different approaches to neural activity underlying behavior can be distinguished: (1) investigations of the nerve message itself, (2) the uncovering of "wiring diagrams" underlying behavior, and (3) electrical activity in the central nervous system.

THE NERVE IMPULSE

Toward the end of our survey period Eccles published his *Neurophysiological Basis of Mind* (1953), and later *The Physiology of Nerve Cells* (1957), which review accumulated evidence about the nature of the nerve impulse and synaptic transmission. Although the details of this fascinating subject take us rather far from behavior, the principal findings are relevant since they tell us the form in which information is transmitted in the nervous system.

The nerve cell at rest maintains an electrical potential difference between its outside and inside. Certain positive ions are accumulated by the cell (e.g., potassium) while others are actively extruded (e.g., sodium). This state of affairs, combined with the electrical effects of negative organic ions trapped in the cell because they are too large to pass through the cell membrane, renders the interior of the mammalian motorneuron about -70mv polarized with respect to the outside. If the neuron's resting potential is decreased by some means, it finally reaches a critical value at which a *nerve impulse* is triggered. The nerve impulse is a massive, rapid inflow of positive sodium ions, which not only depolarizes the cell but also actually makes the interior positive with respect to the exterior. This sodium influx is followed a very short time thereafter by a similar massive outflux of the positive potassium ions, which, of course, tends to repolarize the cell. The sodium influx reaches its peak before the peak outflux of potassium ions so that the potential of the cell rises rapidly to perhaps 20 mv and then declines more slowly, until it actually hyperpolarizes below -70 mv. As potassium outflow slows and the cell begins again to accumulate potassium, the potential returns to the resting value.

This nerve impulse is extremely short, lasting perhaps five milliseconds

(5/1000 sec) from beginning to end. The depolarization of the membrane at one point spreads with decrement to adjacent parts of the membrane, which then similarly become permeable to ion flux. In this way the impulse travels along the nerve fiber (axon). A neuron depolarized in the middle conducts impulses both ways; however, under normal activity nerve impulses travel only in one direction because of the *synapses*, or junctions between cells.

One cell is joined to the next by minute terminal feet implanted on the membrane of the other cell. When a depolarizing impulse reaches these feet, small quantities of a chemical (*neurohumor*) are extruded and diffuse across a minute space (the synaptic cleft) separating the cells. Upon reaching the next cell's membrane, these chemicals somehow alter the membrane's permeability to ions. Some chemicals (such as acetylcholine) cause a depolarization of the electrical potential across the membrane, while others cause a further polarization from the resting state. A depolarization of sufficient strength spreads across the membranal area to the axon hillock, the beginning of the impulse-conducting axonal part of the neuron. Here, if the depolarization is of sufficient strength, it triggers an impulse that travels down the axon to its terminal feet.

It is at the synapse that activity of two previous neurons can summate. Or, if one extrudes a hyperpolarizing chemical and the other a depolarizing one, the first inhibits the second by cancellation.

This is only a very crude summary of the excitation studies on single neurons, but it does demonstrate the nature of the neural message: (1) messages are discrete impulses whose actual electrical level is always the same; (2) the synapse is the place of interaction between previous inputs and also acts as a gate allowing activity to proceed only in one direction. This means that information carried by nerve cells can be only of two principal kinds, the frequency of impulses in one neuron and the temporal-spatial distribution of activity in different neurons. Although some modifying details (see Bullock, 1958, for example, and Horridge, 1968) are receiving recent attention, this general pattern probably holds throughout the nervous systems of most animals.

It is probably a mistake, however, to conclude that only nerve impulses are available as processes for central nervous function. Neurons also display decremental currents, changes in voltage which are below the threshold for the triggering of an impulse. Hence, these are not propagated. This does not mean that they are necessarily unimportant, however. These graded potentials can summate and trigger an impulse. More significantly, the graded potentials may provide a background that represents interactions and activities over a larger area of the nervous system. It could represent an analogue of events, in contrast to the discrete, digital record provided by the nerve impulses (Pribram, 1971).

WIRING DIAGRAMS

An obvious step in analyzing nervous mechanisms of behavior is to seek out, electrically, the connections between neurons. Sir Charles Sherrington made this his life work. His great discovery was of *spinal reflexes*, such as the simple patellar reflex used by doctors as a test of normal neural function. A blow just under the knee stretches a tendon and thus activates a stretch receptor. The receptor sends impulses to the spinal cord, where a single internuncial neuron can transmit the message to motor neurons that cause the muscle to contract. In its simplest form, then, a behavior pattern can function through three nerve cells (an actual reflex uses more since it requires activity in many motor neurons to promote full muscle activity; antagonistic flexor muscles are also inhibited through a parallel channel).

It would be optimistic indeed to assume that all behavior patterns have such a circuit that could be found if only one looked hard enough. Although the knee reflex provides a simple general model for the way in which neural networks underlie behavior, brain histology (above) suggests that most real networks are too complex to be grasped fully by such a model. Further, Sherrington's reflex model ignores the anticipatory aspects of many reflex acts (the "feed-forward" discussed in connection with the von Holst-Mittelstaedt model, Chap. 4; also, see Pribram, 1971). Many sensory cells may be sending information centrally, and this sensory information may be processed by many interacting synapses along the way (the vertebrate visual system has two primary synaptic junctions in the retina itself, for instance). The tremendous possibility for spatial interaction between all the rods and cones of the eye or all the hair cells of the ear make even deciphering of the sensory message for complex behavior an all but impossible task. Then information from various sense systems must also be combined in the brain and must interact with activity states endogenously occurring there to produce a final set of motor commands to the effector organs (glands and muscles). The ultimate goal of all neurophysiology is to decipher these complex wiring systems and to learn how they become active under different stimulus inputs, but it is a goal not attainable merely through use of the methods for studying the patellar reflex.

ELECTRICAL ACTIVITY OF THE CENTRAL NERVOUS SYSTEM

The cephalic elaboration of the spinal cord which we call "brain" is well-known histologically, at least for some animals. It is natural to search for electrical correlates of behavior in histologically differentiated areas, but the method of single-cell recording is too restricted for present techniques since many electrodes would have to be recorded simultaneously in order to obtain an overall picture of activity. However, if a gross-recording electrode is inserted into the brain or is attached to the skull's surface, various slow electrical potentials are recorded under various states of behavior. These potentials are often called *electroencephalographic recordings* (EEGs).

A slow potential recorded from the auditory cortex in response to a sound may be a simple biphasic diversion (that is, hyperpolarization then depolarization with respect to resting level) lasting about 30 msec. It appears only after a certain delay (e.g., 10 msec) after the stimulus. If electrodes are in various parts of the auditory system progressively closer to the ear, this delay is also found to become progressively shorter. Potentials recorded at various stations may be quite different. These potentials evoked by external stimuli are often called *evoked potentials.*

The EEG of a sleeping person is a continuous potential made up of various frequencies of change. Usually, there is a basic rhythm of large phasic changes (several a second) over which is superimposed a number of smaller excursions of higher frequency. An alert person with eyes open gives quite a different record. In this case, the basic slow rhythm ceases and many more frequencies of smaller amplitude are observed. One convenient way to summarize this information is to plot a graph of frequencies in the waves against the percent of time the frequency is present. This allows at a glance the comparison of EEGs taken from animals or persons exhibiting different behavior or experiencing different stimuli.

No one knows for certain the origin of slow potentials. They may be the summed effects of post-synaptic potentials caused by the chemical transmitters (see above) of decremental currents or even the summed effects of the impulses themselves. One interesting thing is that EEG records of animals such as the cat, dog, monkey, or guinea pig closely resemble those of man.

EEG patterns change during learning. If a classical conditional response to a sound is established (cf. Chap. 3), this behavioral change is accompanied by a change in the EEG. For instance, a weak neutral sound produces little change in a certain EEG pattern of slow waves, but a strong light flash depresses large, slow components and evokes fast, small ones. If the sound is repeated prior to the flash, the sound alone comes to depress the slow components of the EEG.

Some slow potentials have been correlated with animal behavior, but their very complexity and obscure origin have prevented any major behavioral phenomena being explained by them. Slow potentials combined with single-cell recordings may, however, be a powerful future tool in the analyses of brain function.

Electrical Stimulation of the Brain

Electrical stimulation of some restricted portions of the surface of the cortex is known to evoke either specific sensory impressions (in humans) or to elicit movements of particular portions of the body. Thus, an isomorphic representation of the body is believed to exist within the confines of the motor and sensory cortex, respectively. A conventional *map* of the cortical areas is shown in Fig. 4–2. Removal of portions of any of these areas leads to the development of a corresponding sensory or motor deficit. This apparent localization of function promoted and supported reflex theories of behavior, with either single cells or minute cortical areas acting as the "switchboard," each being specifically reserved for a particular reflex arc. However, this view was considerably modified by Lashley's brilliant studies

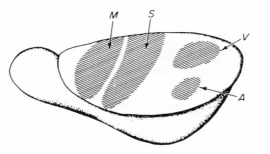

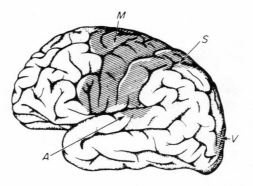

Fig. 4–2. Approximate loci of sensory and motor areas in rat and man. Man's visual area is pushed round into the cleft between the two hemispheres, so that it can hardly be seen. *M*, motor; *S*, somesthetic; *A*, auditory; *V*, visual. (From Hebb, 1949.)

of the effects of brain damage (see below) that led to the belief in more general functioning of the entire cortex in all behavior.

Quite different studies, initiated by W. R. Hess (1949), also employed brain stimulation. Electrodes inserted into certain areas in the brain of cats elicited "sham rage" upon stimulation, thus maintaining the concept of discrete centers responsible for behavior. Other centers served to maintain operant behavior, i.e. electrical stimulation served as a conventional "reward" (Olds, 1958). These centers, however, were located in the hypothalamus, not cortex, so that the question of cortical localization of function was not solved. Since brain stimulation studies have blossomed in recent years (see Some Recent Developments at the end of this chapter), more can be expected from this kind of analysis in the near future.

Electrical Interference in the Cortex

Some workers, for instance, Köhler and Held (1949), have attempted to describe the diffuse (i.e., nonlocalized) nature of the cortex's integrative prowess in terms of *electrical fields*. This work grew out of gestalt psychology. The gestaltists had previously formulated certain "laws" of form (cf. Köhler, 1947). Their laws, originally over a hundred in number, specified the principles according to which objects are perceived (for a critical discussion of these laws, see Allport, 1955). For instance, a circle with a segment missing is perceived as whole; two adjacent lights flashing in succession are perceived as a single, moving light. The laws of perception imply that the perceived form is organized and, to a degree, organized independently of the perceived object. The perceived form, in other words, is a psychologic state. It is assumed that there is an isomorphism or identity in form between the configuration of any perceived form and the perceiving apparatus, i.e., the brain. This isomorphism is seen as an identity in configurations.

In the case of the brain, the configurations are believed electrical in nature. Thus, when an observer examines a picture in which he perceives a separation between figures and their background (as, for example, a dark square on a light background) there will be an area of the cortex in which a region of greater activity is surrounded by a region of lesser activity. The cortical areas will be topologically, if not spatially, identical to the visual field being perceived. The cortical activity gradients that are associated with perceptions, in other words, bear a functional relation to the stimulus pattern. The laws of gestalt perception are thus believed to reflect basic laws of cortical organization which order all stimuli.

As Allport (1955) has commented, the gestaltists have evolved "an elegant theory, beautiful in its logical consistency, and explaining (if true) a large number of facts of perception. . . . The only question that remains is whether it conforms to known facts about the cortex of the brain" (p. 135). In the present context, however, what is most important about the work of Köhler and his colleagues is that its emphasis on the relational properties of stimuli rather than their absolute aspect was also incompatible with earlier views favoring cortical localizations of specific perceptions.

If cortical function is best described in terms of electrical fields that involve all of the cortex, with specific regions expendable insofar as the integrity of the structure of the field as a whole is concerned, interference with these fields should lead to specific perceptual distortions. Lashley, Chow, and Semmes (1951), and later Sperry (1956), attempted a test of this prediction by implanting many small strips of either an insulating or conducting material throughout the cortex. One might expect such treatment to radically alter the electrical field structure of the cortex. In point of fact, all workers were unable to discover any major intellectual impairment as a consequence of these implants. The slight defects that did occur were attributable to actual tissue damage. Other criticisms of field theories by D. O. Hebb are noted below. For the moment, we will leave the controversy between proponents of field and localization theories in the unresolved state intimated by the facts thus far cited.

Neural Destruction and Rearrangement

So far we have considered two primary approaches to brain function: neurology and neurophysiology. By far the greatest contribution to cortical functions made during the first half of the twentieth century was by Lashley in a series of experiments culminating in the publication of his *Brain Mechanisms and Intelligence* (1929). His method of brain destruction might be called an extension of neurological studies, but it merits separate discussion. His studies, and those of experimental rearrangements of neural connections by Weiss (1941), were also aimed primarily at the central question of localization of function (particularly learning) in the nervous system (and note Gaze and Keating, 1972).

NEURAL INJURY

Lashley's technique was to find the effects of brain damage on behavior. He assessed the retention of learned behavior (principally maze running in rats), the ability to learn new behavior, and the general behavior of the animal following brain destruction. Destruction was primarily of two types:

(1) mass removal of large areas, called *ablation*, and (2) severing of connections by transverse cuts across the brain (affectionately known as the "bologna technique").

First of all, Lashley noted considerable variations in the results of cortical ablations upon laboratory rats. Deficient functions (e.g., visual discriminations) would be at least partially restored with time, and the effects of extirpation varied between experienced and naive subjects as well as with the total mass of cortex removed. Equally compelling was his realization that visual discriminations were generally made on the basis of relational rather than absolute cues (see Chap. 3). A rat is first taught to select the larger of a pair of cubes (cube B). Then if an even larger cube (cube C) is substituted for the smallest of the series (cube A), the rat will select the new cube, C. Thus, localization of a precise sensory image (which implies attention to absolute, not relative properties) cannot be involved. Similarly, in the case of conditioned and complex motor response, different muscles or movements may be used to achieve an identical end. A cat that has learned to escape from a Thorndike box (a cage with a latch that must be manipulated in a particular fashion before the cage door will open, cf. Chap. 3) can readily resort to alternative stratagems if the paw it is accustomed to use is immobilized. Again, a specific and localized reflex pathway cannot be involved.

Lashley's alternative proposals have been known under the title of *mass action and equipotentiality*. Briefly considered, these terms were intended to signify the equal capacities of different cortical portions and their ability to substitute for one another. Thus, the total sensory, motor, or intellectual deficit upon cortical extirpation would depend on the total mass of cortex removed. This does not deny the existence of some specialized (and localized) cortical areas, such as Broca's speech center or the retinal projections to the striate cortex. However, Lashley emphasized that areas specialized for one function may also be important (in a specific or nonspecific fashion) for other functions as well. Extirpation may truly lead to a specific sensory or motor deficit, but a more general intellectual deficit will appear only as an ever increasing area of the cortex is invaded and removed.

Loss of function, it should also be noted, may be expressed in terms of ease of arousal or degree of organization, with lesions either (1) reducing the readiness of an organism to respond or to learn or (2) actually abolishing previously learnt discriminations. In the latter case, the interference may be with complex though not with simple tasks, which also suggests the overlapping levels at which the cortical areas function.

Equipotentiality of function, in which neither particular perceptions nor intellectual functions are specifically localized, implies cortical self-regulation. Thus, Lashley (in Beach et al. 1960) cites the case of the patient blind in half of each eye (hemaniopsia) whose visual center of fixation had shifted to the periphery of the retina. The center of fixation is generally a region of maximum density of retinal cells (the macula lutea). The new center possessed a visual acuity exceeding that of the macula even though it presumably had a much lower density of retinal cells. Neurons must be continually modifying one another in order to account for such results! An alternate way of viewing equipotentiality is discussed under the section on recent developments below.

EXPERIMENTAL REARRANGEMENT

Additional evidence relating to cortical localization came from embryological studies by P. Weiss (1939), who showed that supernumerary limbs can be grafted onto salamanders. Nerve fibers grow out to the graft in an apparently random manner, similar to the random proliferation of fibers that occurs when an amputated limb regenerates. Once functional connections are established, however, whatever the point of origin of the fibers, properly co-ordinated movements can occur. In some, as yet unknown, manner the nerves are "named" according to the organs they innervate. Thus, their cortical representation can hardly be fixed and invariant. These facts then led to rejection of the reflex and localization hypothesis.

With this brief survey of methods for analyzing brain mechanisms we leave the question of specific localization of function. In sum, it could be said that however wiring diagrams underlying behavior are organized, some considerable overlap of function occurs in the connections of the mammalian cortex. Future studies will certainly combine many of the methods outlined here (histological differentiation, electrical recording, electrical stimulation, brain damage) in order to further uncover brain mechanisms.

Chemical Studies

It should be apparent that the rubrics dividing this chapter are interposed arbitrarily. Studies of the chemical changes within the brain have the same goals and often the same design as those utilizing electrical measures or surgical intervention. The specialized skills that biochemical studies demand, however, have led to the chemical approach to brain function being less commonly employed, and, in addition, it is the most recent of the general techniques utilized for central nervous analysis.

The first chemical studies sought merely to correlate the presence of particular compounds with the development of sensory functions. Thus, in the developing amphibian embryo the electrical responsiveness of the tectum to peripheral stimulation with light was correlated with a rise in acetylcholinesterox concentrations (ACh being involved in synaptic transmission, see below). The major impetus for chemical studies, however, arose from the belief that macromolecules could provide a substrate for information storage (e.g., memory), in a manner analogous to the DNA of the chromosomes. It was thus not entirely *coincidental* that the nucleic acids, especially RNA, should elicit serious attention.

A detailed account of studies of RNA (and other macromolecules) and their role in behavior is provided by John (1967). The kinds of approaches employed include studies of RNA synthesis and turnover in particular sites as a function of learning, or correlated changes in base ratios; interference with RNA synthesis; transfer of brain RNA from "trained" to "untrained" animals in the expectation that the latter would then train more easily. The experiments have been numerous, complex, and often defying replication. John's own conclusion, which it is still difficult to dispute, is that " . . . none of the experimental results . . . is sufficient by itself to warrant the conclusion that stable information storage in the brain is mediated by the structure of macromolecules. Yet, it must be conceded that a wide variety of experimental procedures have yielded an impressive quantity of data strongly pointing towards RNA and protein synthesis as deeply implicated in the functions of memory" (John, ibid., p. 124). Lest this seem a prejudiced position it would be added that John was among the first workers to focus on the role of RNA. Together with Corning (Corning and John, 1961), he analyzed learning in a flatworm, *Planaria dugesia*, and implicated ribose nucleic acid (RNA) in memory functions. Normally, *Planaria* show cephalic dominance. Also, both head and tail sections will regenerate whole individuals. If a conditional response is established in an animal and the flatworm is then cut in two, both the regenerants derived from head and from tail sections show a retention of the response. If the regeneration takes place in a medium to which some ribonuclease has been added, an enzyme that degrades RNA, then only the regenerants from head pieces show retention. Corning and John concluded that the presence of an RNA code in the dominant head-end is essential to the retention of a learned trait. The RNAse interfered with code replication in the regenerants. However, the head-ends already possessed a store of coded RNA and thus the presence of RNAse, while interfering with coding in the regenerated tail, did not interfere with retention itself. The tail-ends, on the other hand, developed new heads which, because of the RNAse, were devoid of the code. Cephalic dominance suppresses any effect on the response the tail might otherwise exercise. The mechanism of RNA coding, if any there be, still remains to be suggested.

Brain Functions

The previous discussions have pointed out the close involvement of the cortex and other brain mechanisms in learned behavior. Chapters 2 and 3 have also raised briefly the question of internal states of activity (drives, motivation, etc.) which presumably occur in the brain. These two functions of the central nervous system will now be considered further.

NEURAL CORRELATES OF LEARNING AND PERCEPTION

If the central nervous system is the locus of learning and perception, then specific physical changes must occur as the valence or meaning of particular stimuli undergoes change, differentiation, or generalization. The first explicit models of neural changes upon learning were based on the assumption of linear conduction of impulses (along nerves) and at least a modicum of cortical localization. The belief was that once a train of impulses had traversed a particular path, the synapses along that path were in some way altered so as to facilitate the subsequent use of that same pathway. It was not difficult to see how this could allow for the various kinds of conditioning that were known or to postulate the nature of the facilitory changes (cf. Eccles, 1953: 220, Fig. 4–3, and 1964). Any alteration of structure that could increase the probability of or reduce the delay in transmission imposed by the synapses response threshold would do. Eccles (cf. 1964) originally proposed a change in the volume of the synaptic end bulbs. Upon passage of an impulse, changes in the ionic concentrations across the neural membrane are known to occur which, in turn, alter the membrane's selective permeability. Permeability changes, by altering rates of water movement, could allow for volume changes that would alter the physical or topological relations vis à vis adjacent elements. Once an alteration occurred that reduced the threshold for the passage of a second impulse, this second and subsequent impulse would have a progressively increased probability of passage. A permanently facilitated or sensitized transmission circuit would thus be assured. Others (e.g., Ariëns Kappers et al., 1936) have suggested a growth in the number, linear extent, or size of the synaptic end bulbs as a consequence of stimulation, although conclusive evidence for such changes is lacking. A specific chemical model was advanced by Thesleff (1962) and will illustrate possible mechanisms (see Fig. 4–3).

According to Thesleff (1962), embryonic skeletal muscle cells are uniformly sensitive to the motor nerve's chemo-transmitter. Once a functional synapse is established, however, the extent of the chemo-sensitive area

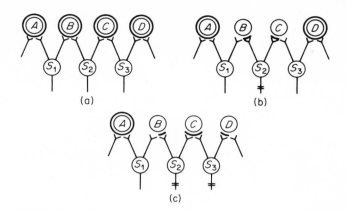

Fig. 4–3. Thesleff's hypothesis concerning the synaptic basis of learning. A, B, C, and D represent neurones which are innervated by three other neurones, S_1, S_2, and S_3. In (a), none of the S-neurones has been stimulated, and consequently the A-D neurones are uniformly sensitive to the transmitter substance. In (b), neurone S_2 has fired, and thus neurones B and C now have developed a localized sensitivity which restricts their excitability to impulses from S_2. These neurones are now "set" to respond exclusively to a specific input. In (c), S_2 and S_3 have fired in an overlapping time sequence. As a consequence, neurones B and D are "set" to respond exclusively to impulses from S_2 and S_3, respectively, while C will respond to *either* S_2 or S_3. The mechanism for the establishment of a conditional reflex has thus been provided. (After Klopfer, 1962; based on Thesleff, 1962.)

decreases. Thus, neurons become set to respond preferentially to inputs arriving at a particular locus and thus from particular axons. In this manner specific nerve networks may be built up and maintained.

The development of specific nerve networks that maintain themselves through a continuous facilitation does imply, of course, that there is a specific locus for memory traces, even though this locus may be spread throughout the cortex. Thus, it might appear to run counter to the interpretations by Lashley of his extirpation experiments which he believed to speak against localization theories. Evidences from the study of image constancy were also adduced against anatomical localization of memory (by the gestaltists, e.g., Köhler, 1947). Yet it is hard to conceive of a memory trace that is not basically structural. The diffúse, nonlocalized, and shifting fields of electrical activity, which code information in the patterns they form, as described by Köhler, simply do not provide a model by which one can understand the permanence of memory traces. Nor, in fact, does the temporary interruption of the cortex's electrical activity, by

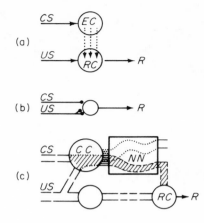

Fig. 4–4. Diagrams illustrating attempts to explain conditioned reflexes in terms of plastic changes in synaptic connections: *CS* and *US* show input into central nervous system of conditioned and unconditioned stimuli, respectively. *EC* represents the emitting center, *RC* the receiving center. The dotted lines between these two centers indicate potential connections. In (a) the arrows indicate nervous pathways, while (b) is a redrawing of a highly simplified model showing converging synaptic connections of the *CS* and *US* lines. The one synaptic knob on the *CS* line indicates that any impulse in that afferent fiber is normally inadequate, while an impulse in the *US* line is assumed adequate to excite a discharge from the neurone, evoking the response *R*. In (c) nervous pathways are drawn as broad bands along which conduction occurs as explained in Eccles, Fig. 76 and text pp. 221–226, particularly in the neuronal network, *NN*. The interruptions in the bands indicate synaptic relays. Nerve centers containing large populations of neurones are indicated by circles, while the neuronal network, *NN*, would be an extremely complex neuronal system, for example, an area or areas of the cerebral cortex. (After Eccles, 1953.)

electro-shock for example, abolish long-term memory, although it may influence recollections of recent events. Somewhere the *engram* must reside, by which a structural trace is implied, despite the experimental evidence that appears to deny this, too.

A partial reconciliation is afforded by the brilliant synthesis of D. O. Hebb's (1949) *Organization of Behavior*. Hebb points out that, in fact, Lashley's extirpation experiments are not inconsistent with the notion of a structural, though diffuse, trace. What Hebb proposes is that the repeated stimulation of specific receptors leads to the facilitation of co-ordinated action by an *assembly* of association-area neurons. These form reverberatory networks, or closed systems, which may continue to fire with the cessation of the original stimulus. This model does assume some alterations at the synapse. It differs in an important respect from previously proposed models in that single cells are not considered capable of establishing transmission networks. The cell assemblies which go into any closed system may be spread over an extensive region of the cortex, and any individual cell may belong to several or many different assemblies.

Hebb's model implies a number of characteristics whose existence is in dispute. First of all, the model implies the absence of most (if not all) built-in perceptual mechanisms: the cell assemblies develop in a trial-and-error fashion. In support of this implication Hebb points to data suggesting that even the simplest visual perceptions require learning. Congenitally blind persons, upon removal of cataracts and the restoration of vision, have great difficulty recognizing shapes or discriminating between circles and squares, though a discrimination by touch is made readily. However, the existence of seemingly innate perceptual mechanism in other forms than man is considered an article of faith among many ethologists (Chap. 2). Perhaps innate perceptual mechanisms are due to peripheral sensory filters which enhance particular stimulus configurations (see Chap. 11). On the other hand, Hebb may be mistaken and the perceptual deficiencies of his cataract patients may be due not to inexperience but to physical deterioration of retinal cells or neurons in the visual pathway. It is also unnecessary to deprive the brain of all "unlearned" cell assemblies. That is to say, natural selection could well place so great a premium on particular patterns of neural activity that certain synapses are facilitated in advance of external stimulation. These "pre-set" circuits would form the basis for what have been termed innate perceptions. (See also the discussion of research by Hubel and Wiesel at the conclusion of Chap. 6 and in Chap. 11.)

Second, as Hebb goes on to suggest, the doctrine of equipotentiality of cortical association areas is not based on the firmest of evidence. Experiments on pattern perception of animals subjected to varying degrees of cortical loss have provided most of the data for equipotentiality. These experiments, however, have neglected both eye movements and prior experience. If one makes assumptions opposite to those of Lashley and Köhler, i.e., assumes extreme cortical localization with a point to point projection of the retina onto the striate cortex, for example, appropriate eye movements (in experienced animals) could still provide for pattern recognition. Lashley and Köhler also assume that the perceptions are unitary, but Hebb's experiments suggest that they are serial, at least in inexperienced organisms. The congenitally blind, upon restoration of their sight, must at first count corners in order to differentiate squares from triangles! The whole pattern is not perceived at once but is built up from its component parts, although with experienced animals the rapidity of the process may create an illusion of perception of "wholes." The fascinating studies of visual and tactile discrimination in octopus (recently summarized by Wells, 1962), though also outside the scope of this study, must be mentioned as having relevance and lending additional force to Hebb's conjectures (Wells' work is discussed below).

In fairness to Hebb, it should be pointed out that he has continued to revise and criticize his own theory. For instance, he (1961) pointed out that the theory suffered from vagueness in identifying the locus and size of cell assemblies and how "separateness" of systems is to be defined. Hebb also feels his treatment of emotion presents difficulties surmountable only by a real restatement of how assemblies work. Finally, newer physiological facts (e.g., cellular inhibition, time patterns of neuronal firing, and the activating role of the reticular formation) necessitate a major revision of the original theory. Nonetheless, Hebb's (1949) theory remained for years a general guideline to problems of the mechanisms of learning. (A major successor to it is described under Recent Developments.)

It appears, then, that learning involves physical changes of one sort or another at the synapse. However, the networks that are activated upon the stimulation of sensory cells do not involve linear transmission lines but large cell assemblies, firing in a co-ordinated fashion. They also involve established reverberatory, self-exciting circuits. These are spread over extensive regions of the cortex with any one cell being both party to many assemblies and not indispensable to any assembly.

Among the most striking of recent studies which relate brain structure and function are those of Wells (1962), Sutherland (1960), J. Z. Young (1961) and Boycott (1954), all of whom have exploited certain of the extraordinary traits of *Octopus*.

Octopus vulgaris will readily adjust to a life in a laboratory aquarium, a pile of bricks in one corner of the tank serving as home. From this home, the beast will make periodic forays to capture whatever prey organisms the experimenter chooses to introduce, e.g., small crabs. By pairing crabs with one of two symbols painted on a card, and pairing an electric shock, delivered by means of a cattle probe, with the other symbol, an octopus will rapidly learn to discriminate between symbols. One symbol will then initiate an attack, the other a retreat. Similarly, blinded animals may be trained to respond discriminatively to plastic cylinders whose surface is scored in different ways. These simple responses have served as the basis for elegant analysis of brain function.

Turning to the problem of visual discriminations first, let us consider the portions of the brain serving the visual sense. It has been shown by J. Z. Young (1961) that the optic lobes of octopus contain dendritic fields whose primary axes lie at right angles to one another. Suppose these dendritic fields represent a symmetrical grid. Then the projection of two-dimensional figures upon this grid will be represented by differences in the number of horizontal or vertical cells of the grid that are engaged. Figure 4–5 illustrates the results of having various shapes superposed onto

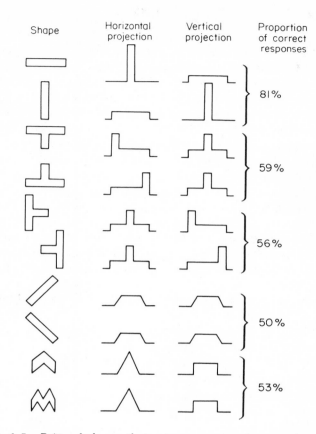

Fig. 4–5. Pairs of shapes that octopuses can and cannot learn to distinguish. Percentage figures show the proportion of correct responses made by groups of octopuses in the first 60 trials (pairs 1–4) or 240 trials (pair 5) of training. (After Wells, 1962.)

such a grid, the horizontal and vertical cells of which would be covered by the particular shape in question. Additional experiments have shown that horizontal projections are less easily discriminated than vertical projections, a fact that accords well with the observation that in the optic lobes vertically oriented dendrites exceed horizontally oriented ones in number (as cited by Wells, 1962).

Tactile discriminations as made by sightless animals appear to depend on the number of adjacent tactile receptors that are stimulated, i.e., on the proportions of grooves on the object and not on the orientation or pattern of grooves. Thus, horizontally and vertically grooved cylinders will not be distinguished, provided the dimensions of the grooves are similar. Variations in the extent or width of grooving are readily detected, however, irrespective of visually detectable similarities in groove structure. As in the case of the visual system, the nerve projections from the sensory endings of the arms to the brain retain a spatial organization consistent with this presumed mode of discrimination. It is hard to imagine a more striking case of isomorphism with respect to the primary sensory fields, the objects that initiated their stimulation, and the structural organization of the central nervous system itself.

This same animal, *Octopus vulgaris*, has also provided some evidence directed to the question of central integrations. Wells (1959) found that in tactile discrimination tests where trials were spaced but 3 to 5 minutes apart, the performance of one arm showed no transfer to a second arm when it was tested. When the same number of trials were spaced 20 minutes apart, however, transfer from one arm to another did occur. Thus, it appears that each arm is represented by a functionally distinct neuronal field within the central nervous system. Passage of information from one of these fields to another requires an interval of time of more than 5 though less than 20 minutes.

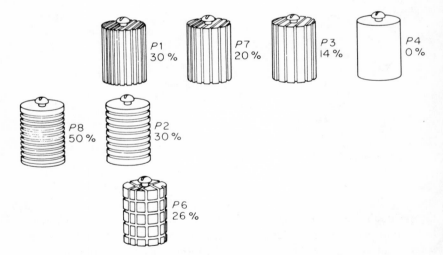

Fig. 4–6. Perspex cylinders used as test objects in tactile training experiments. The series is arranged so that there is a decreasing proportion (given as a percentage beside each object) of grooved to flat surface from left to right. The top row is of cylinders that differ only in the frequency of standard sized vertical grooves cut into them. P_1, P_2, and P_6 differ in the arrangement of the grooves, which represent the same or almost the same proportion of the total surface in each case. Octopuses find these three difficult or impossible to distinguish but can readily be taught to discriminate between the members of the top row. (After Wells, 1962.)

THE ACTIVATING ROLE OF THE
CENTRAL NERVOUS SYSTEM

Behavior is not mere reaction to externally imposed stimuli. This fact has long been recognized, at least implicitly. The concepts variously known as "set," "stimulus trace," "central motive state," "drives," etc., attest to the existence of autonomous central processes.

Direct evidence for the autonomous activity of the central nervous system can be obtained both by electroencephalographic and direct recording from implanted electrodes. Resting, deafferented, and even isolated slices of brain all produce waves of electric activity. In whole, deafferented, or resting brains, the EEG patterns show synchrony and hyperpolarization. Upon sensory stimulation, the synchrony is disturbed and the magnitude of the changes in potential reduced (see above).

On a more gross level, reactions of organisms may be considered as responses to specific stimuli or to a deficit. The courtship behavior of a bird that has long been isolated from its kind, a so-called *vacuum activity* (cf. Chap. 2) would be an example of the latter. These reactions in their apparent independence of external input also suggest an autonomous activity of the centrum.

One of the most striking illustrations of the manner in which central nervous activity may dominate peripheral (sensory or motor) events, was provided by von Holst and Mittelstaedt (1950; also cf. von Holst, 1954 summary) in their *reafference theory*. Studies of brain function have classically been based on the query, "What is the effect of a particular input on central activity?" Von Holst, reversing the question, asked what effect has the central nervous system on the peripheral system? To illustrate the significance of this second form of the question, assume that you are standing quietly with your arm against the bough of an apple tree. In the first instance, you decide to shake the bough, whereupon a motor impulse (the *efference*) causes certain muscles to undergo periodic contraction and relaxation. Information as to the state of each muscle is relayed back to the brain from proprioceptors (the *afference*). In the second instance, while you are standing motionless, a breeze caught by the bough moves your arm. Assume that the identical muscles are moved and contracted to the same degree and in identical temporal sequence as in the first instance. Again, a proprioceptive input, or afference, is provided. In this latter instance, however, the ultimate source of the input was an agent external to yourself, the wind. The afference produced by the action of the external environment can be labeled *ex-afference*, that resulting from voluntary movement, *re-afference*. The temporal and spatial pattern of proprioceptive impulses may be identical in both cases. Nonetheless, our perceptual apparatus distinguishes between the two kinds of afference: normally there is no question of whether you are voluntarily moving your arms or whether they are being moved by an external source. Thus, the central nervous system imposes a meaning on the reafference different from that imposed upon the exafference. This, in turn, is evidence of central activity that proceeds independently of the periphery.

Von Holst's model assumes that every efference leaves an "image" of itself at some lower center to which the reafference must compare as a negative to a positive print (Fig. 4–7). Superposition of the negative and positive may erase the image, or *efference copy*, but should erasure not occur, predictable illusions may follow. Thus, if your eye is mechanically fixed with the muscle proprioceptors narcotized and you command it to turn to the right (i.e., from an efference copy), the visual field will appear to jump to the right. No reafference could result, of course. Now, let your eye be turned to the right mechanically. This time the field appears to move to the left. Movement of images across your retina produced a simulated reafference but there was no efference or efference copy. If these two operations are simultaneously executed, i.e., your eyes turned right mechanically just as you command them to do so, the visual field appears stationary!

The question of the dependence of central activity on prior experience remains. Hebb attributes autonomous activity to the stable patterns of activity of large neuron assemblies. Motivation, interest, set, and other manifestations of autonomy would be due to organized and persistent se-

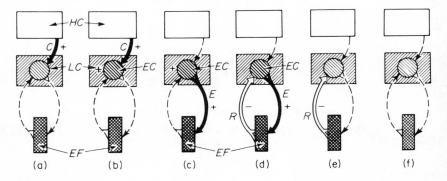

Fig. 4–7. Diagram of the reafference principle. (a) A "command," C, from a higher center, HC, causes a certain activation (b) in a lower center, LC, establishing an "efference copy," EC, which (c) produces an efference or motor impulse, E, to the effector, EF. The efference produces a motor response (d-f) whose proprioceptive feedback or reafference, R, returns to LC and erases EC by superposition. Because of the complementary action involved, the efferent component is designated as +, dark areas, and the afferent component is designated as −, white areas. (After von Holst, 1954.)

quences of patterned activity. These sequences, however, might either arise gradually during the lifetime of an individual in response to external stimuli or might develop entirely independently of an input. The totality of the latter case would doubtless prove fatal to the organism. By the same token, the ubiquity of external stimuli even during the earliest stages of development would assure the presence of at least some reverberating circuits.

Maturational Changes in the Central Nervous System

The ontogeny of the vertebrate central nervous system has been repeatedly described by embryologists (cf. Weiss, 1939). Thus, it has not occasioned surprise when it has been assumed that age-dependent changes in behavior are related to developmental processes affecting brain structure or function. The point at which arguments begin is where one must consider whether such developmental processes affecting brain structure or function proceed autonomously or whether they are dependent on experience or sensory input, i.e., it is a restatement of the old competition "nature vs. nurture."

In some respects, of course, this question misrepresents the situation in which real, live organisms find themselves as distinct from that of the rhetorical creatures that stalk the pages of many articles on behavior. Complete sensory isolation is probably only an experimental artifact. Even an embryo within the maternal uterus receives much and varied stimulation. Thus, given a genetically imposed degree of constancy of the environment (intra-uterine) in which a group of organisms develop, the nature-nurture quarrel becomes largely a semantic battle. If the conditions necessary for normal development are unchanged and unchanging through successive generations, then, however much effect an input has on the maturing nervous system, its development proceeds as if autonomous.

Despite the intimacy of the relation between the developing brain and the environment external to it, examination of the precise nature of the interaction is important. One of the earlier of such studies is that by Preyer in 1885 (translated and published in English in 1937). His detailed descriptions of the movements and sensitivity of mammalian embryos of a wide range of ages led to an important generalization: motility precedes sensibility. Although some workers, e.g., Carmichael (1926), found this supporting their own conclusions on the independence of maturational changes from sensory stimulation, it can be shown that a different interpretation follows as readily. The measure of sensitivity is an overt motor response to a stimulus. The failure of a stimulus to evoke a response can, however, be due to the absence of sensory-motor connections alone. It need not be due to an absence of sensory input to the centrum. Thus, one may suggest the existence of a feedback system such that an input to the brain triggers developmental changes or responses that, in turn, modify the subsequent input.

Recently, for example, Shen (1953) demonstrated that the appearance and distribution of acetylcholinesterase in the optic tectum of amphibia appeared to parallel the development of functional synapses in the visual system. From the other studies we know that the failure to stimulate retinal cells causes some central degeneration (Weiskrantz, 1958). It might well be that a reciprocal relation exists between, let us say, acetylcholinesterase production and retinal stimulation. For the moment, of course, this is but conjecture.

A good many experiments have shown that development of particular behavior patterns is unrelated to actual functioning. Amphibian larvae maintained under anesthesia and birds reared in restricting cylinders all show normal or near-normal locomotor movements at the appropriate age when released from their restraints (Weiss, 1939; Carmichael, 1926). Depth perception may also be evident in rats reared in darkness until the time of their test (Walk and Gibson, 1961). However, this only states that the performance of a particular task is not essential to the development of the neural capacities essential to that task. It does not by any means transform the developing nervous system into an autonomous organ removed from and independent of the sources of external stimulation (and refer also to Chap. 2).

Some Recent Developments

Few areas are attracting so much attention as the interface of nerves and behavior. Pronouncements as to the "current" view or trend are almost guaranteed obsolescence before the ink has dried. Yet, we can perhaps identify two thickets which seem likely to be actively explored for another decade. These are represented by the questions: What is the basis for memory? Does the central nervous system influence perception; and if it does, how?

MEMORY

In an earlier section we contrasted theories that localized memory stores in a particular region with those that sought the store in stable reverberating patterns of electrical activity. The behavioral phenomenon of memory, however, is hardly a discrete process. At the least it involves one or more transitory or "holding" stages, a period during which consolidation or permanent encoding of the transitory events takes place, and once a "permanent" memory has been recorded, there must be a mechanism for "recalling" that record (John, 1967). Even for any one of these events, however, there need not be a single mechanism. "Neural modifiability is multifaceted, and memory is *not* a unitary process," writes Pribram (1971), a leading student of the subject. He proposes an embryological model, one based on the phenomenon of induction, in which certain substances (the "inducers") derepress processes capable of occurring within certain substrates. This phenomenon is also analogous to derepression of the genetic mechanism in the earliest stages of development. Pribram describes the inductor-substrate relationship and its applicability to the processes of memory as follows:

1. Inductors evoke and organize the genetic potential of the organism.
2. Inductors are relatively specific as to the character they evoke but are relatively nonspecific relative to individuals and tissues.
3. Inductors determine the broad outlines of the induced character; details are specified by the action of the substrate.
4. Inductors do not just trigger development; they are a special class of stimuli.
5. Inductors must be in contact with their substrate in order to be effective; however, mere contact is insufficient to produce the effect—the tissue must be ready, must be competent to react.
6. Induction usually proceeds by a two-way interaction, by a chemical conversation between inductor and substrate.

Evidence of the role of RNA in memory storage can at present be suggestively explained by recourse to a model based on this embryo-genetic process of induction. The model states that excitation of nerves is accompanied by RNA production. This neural RNA induces changes in the surrounding oligodendroglial cell commencing a chemical conversation indicated by the reciprocal nature of the variations in concentration of RNA (and a host of metabolites) between neuron and glia. A change is induced in the functional interaction of the glial-neural couplet. This change may in the first instance induce comparable RNA in the glial cell which then over a longer period of time produces alterations in the conformations of lipids, proteins, and lipoproteins, all large molecules which make up the membranes interfacing neuron with glia. Such macromolecular changes can alter the ease with which chemical neurotransmitters are released or destroyed. These configurational changes are reversible and can fade or be superseded. When maintained by repetition of the same pattern, however, the alterations in molecular conformation will endure long enough to produce an effective change in membrane permeability which, in turn, allows more RNA, metabolites, and neurotransmitters originating in the excited neuron to affect its glial surround to the point where glial cell division is actually induced. Once divested of its encapsulating glia, the growth cone of the neuron is free to plunge between the newly formed glial daughter cells and to make new contacts with the neurons beyond. Thus, the induced cell division of oligodendroglia is assumed to guide the growth cones of central nervous system neurons much as those in the periphery are guided by the related Schwann cells (Pribram, ibid., pp. 42–43).

The paradox of memory traces which are both localized (i.e., can be elicited by points stimulation) and diffuse (i.e., shows persistence even after massive cortical extirpation) is resolved by analogy with holography. In the optical holograph, most of the original scene can be reproduced from but a fragment of whole, though with some loss of resolution or depth of field. The changes reflected in the brain as a consequence of experience could be coded in an analogous manner. Perhaps this is the significance of the decremental potentials.

The compatibility of this model with the experimental facts as detailed by Deutsch (1960), Milner (1960), and John (1967), Horridge (1968), and Honig and James (1971) has yet to be systematically explored. However, it does provide a theoretic framework within whose compass specific, testable questions can once again be asked.

THE ACTIVATING ROLE OF THE CENTRAL NERVOUS SYSTEM

The concept of the central nervous system as a receiver, a passive device dependent on inputs from sense organs, has not been acceptable for some considerable time. The degree to which the central nervous system actually tunes or biases the input channels, thus directly influencing perceptions at their source, is only gradually coming to be appreciated. Perhaps the best illustration of this aspect of central nervous function is a model of emotion developed by Pribram and Melges (1969). The important feature of the model in the present context is the stress it places on the initiating role of the brain.

Consider an experiment by Sokolov (1963). A tone of fixed duration and intensity is regularly presented to a subject from whom a variety of measures of "alertness" or "orientation" are taken—electroencephalogram, galvanic skin response, etc. After repeated presentations, the changes in these measures which the tone originally induced become smaller, finally disappearing. The subject has "habituated," just as city dwellers cease hearing traffic noises outside their windows, or farmers their matudinal cocks.

If the tone is then *reduced* in intensity, or even eliminated, the earlier responses reappear full-blown. Apparently, the nervous system can measure discrepancies between some established pattern of activity and some change which has the effect of altering the pre-existing configuration of events.

Pribram and Melges (1969) suggest emotions reflect two kinds of processes: those that reflect the state of disorganization of ordinarily stable configurations ("concurrent emotions") and those that reflect those mechanisms by which an imbalance is redressed by means of a regulation of inputs ("prospective emotions"). Figure 4–8 illustrates the system.

The input represents any kind of stimulation to which the organism is subjected. Over a period of time, this input becomes associated with a particular configuration of neural, hormonal, and visceral activity. This is the baseline. When that particular pattern of stimulation recurs, the pattern that initially established the baseline, it is subjected to a test. If the input produces the same configuration, if the result is "congruity," matters end there. The process that is involved in the confirmation of a congruity represents a concurrent emotion.

If the input alters from that which is expected, arousal results. This requires an appraisal of the significance of the mismatch. If the incongruous input is nonetheless a desirable or relevant one, it will be incorporated into the test module and the template will be altered or the baseline changed. This will then lead to a congruity-match. If the input is adjudged as irrelevant or undesirable, the input is filtered or eliminated. This particular phase of the test process gives rise to the prospective emotions, optimism or pessimism, depending on whether the appraisal of the incongruous input is that it is relevant or irrelevant.

Incorporation involves changing the neural model. Filtering the input

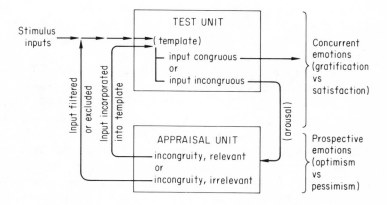

Fig. 4–8. A model of emotion. (After Pribram and Melges, 1969.)

or gating in its extreme form may give rise to the condition known as *narcolepsy*, in which individuals go to sleep in the face of stress. This blocks all input and is certainly the most effective form of gating. If gating is unsuccessful, positive feedback may ensue. This leads to a progressive divorce from reality. Gating, in short, represents an adaptation which, when carried to extremes, produces hyperstability and inflexibility.

Incorporation, however, must not be thought of as the solution to any incongruous input. The individual who readjusts his baseline with each incongruity will be unstable and hyperreactive.

There are other interesting implications of the model that increase its credibility. Consider again the effect of gating on the perception of the passage of time. Gating reduces the information content that reaches the test center. It, therefore, enhances redundancy, and the speed with which the internal processing occurs must, therefore, speed up. That means internal activities become fast relative to external activities. There are two clocks, an internal clock and an external clock. The latter keeps an absolute rate of time; the internal one is sped up. Then, with respect to the external clock, time is seen to go more slowly; time drags.

Incorporation, on the other hand, implies an increase in the amount of information being processed and a reduction in redundancy which must result in the neural flow. With internal activity flowing slowly relative to external time, time must race on. Thus, the character of the inner world is not dependent simply on external outputs. The brain shapes its own inputs.

Overview

Studies of brain function as related to brain structure exhibit clearly the oscillations common to much of science. During the early nineteenth-century, notions of *cortical localization* were advanced by the doctrines of phrenology, and later by histologic analyses. The functional studies of the neurosurgeons lent even greater support to localization theories, as did Müller's doctrine of *specific nerve energies*. Then came the reaction, in the form of Lashley's hypothesis of *mass action and equipotentiality*, and the studies of the relational qualities of perception of the gestalt psychologists, notably Köhler. Once again, a reaction occurred, occasioned this time by the work of Sperry, who failed to disrupt brain function when the electrical fields of the cortex were (presumably) distorted. He was supported by the elegant model of brain function constructed by Hebb. Hebb's notions regarding *cell assemblies* which act as reverberatory circuits account both for memory and the persistence of percepts as well as for learning. His model implies localization of function, but it is localization within assemblies of many hundreds or thousands of cells so that much in the way of *equipotentiality* can be accounted for. The basis of the stability of these cell assemblies, however, is facilitation of transmission at specific synapses, hence specific localization continues to be an important, if not central, doctrine.

Selected Readings

The most complete compendium on the anatomy of the central nervous system of vertebrates is Ariëns Kappers, Huber, and Crosby's (1936) three-volume tome. A short historical survey is to be found in Boring (1957). Stevens' (1951) handbook has some useful summaries of the status of neural and sensory physiology as well. Hebb's (1949) important book should not be neglected by any serious student of central nervous system functions. Current reviews may be sought in the *Annual Reviews of Psychology*, and *Annual Reviews of Physiology*. Finally, the synthesis by Pribram (1971) is an original and provocative treatment of many of the theories discussed in this chapter.

<table>
<tr><td>Chapter 5</td><td>

HORMONES
AND
BEHAVIOR

The Relations between
Hormonal States and Behavior

</td></tr>
</table>

Introduction

The first half of this century witnessed a change in thinking about hormones and their effect on behavior. During most of this period, hormones were looked upon as isolated chemicals of the body that were released by special internal organs and had specific effects on other internal organs, including the brain. But during the end of our survey period, a new trend in hormonal research appeared, a trend to be discussed at the conclusion of the chapter.

The old view was involved with the question, "What hormone produces which behavior pattern?" Two rather separate groups of "ethologists" studied hormonal effects of behavior. On the one hand, G. K. Noble, Frank Beach, and similarly minded physiological psychologists carried out experiments on reproductive and parental responses in a variety of animals. On the other hand, William Rowan and other ornithologists tried to unravel the role of hormones in avian migration. Although common ground between the groups existed in the study of annual cycles, they followed essentially different paths.

Excluded from consideration in this discussion are general studies in endocrinology, a subject which defies succinct summary. The cellular mechanisms of hormonal action are nearly unknown (but, cf. Schneiderman and Gilbert, 1964; and Hamburgh, 1971), while the effects of hormones on gross anatomical structures, although known, have been made relevant to behavior in only a few cases.

Major Hormones and Methods of Study

Before considering the actual studies of hormonal effects on behavior, let us take a brief look at the major hormones and the methods employed for studying their behavioral effects.

THE MAJOR VERTEBRATE HORMONES

Hormones are specific chemicals, minute amounts of which are released into the bloodstream by endocrine glands. The liberated hormone is taken up at some *target organ*, which may even be another gland. Some hormones act very generally, having no specific target organ, while others have several target organs on which they produce different effects. Some hormones have been analyzed chemically; they are heterogeneous in structure, although certain chemically related groups can be separated (sex steroids, catachol amines, etc.). Hormones certainly vary somewhat from species to species, but there is a surprising trans-specific effectiveness. Mammalian sex hormones, for instance, promote development of gonads when injected into birds and amphibians.

Most of the hormones directly affecting behavior—that is, those to which we shall pay most attention—come from three organs: the pituitary, adrenal gland, and reproductive organs. The pituitary, however, is really composed both of neural tissues (the *neurohypophysis*) and endocrine tissue (the *adenohypophysis*) which produce distinct hormones. The reproductive organs secrete hormones from different sites. For instance, males have only testes; females have ovaries and, during pregnancy, the placenta, for hormone secretion. Furthermore, the corpus luteum, formed when an egg has been ovulated from the ovary, secretes a hormone distinct from the steroids secreted by the ovary *per se*. A brief summary of the physiological effects of the principal hormones secreted by these organs is given in Table 5–1.

It is apparent from inspection of the table that some hormones (e.g., certain steroids) are secreted from several different sites (adrenal cortex, gonads) and that certain sites (e.g., adenohypophysis) secrete a variety of distinct hormones. Some hormones have regulating effects, maintaining a certain concentration of substance (e.g., sodium) in the body, or a certain physical state (e.g., blood pressure), or a certain process (e.g., carbohydrate metabolism). Other hormones stimulate the growth or development of certain organs (e.g., testis), or trigger a certain action (e.g., ovulation). Finally, some hormones act on other endocrine glands, stimulating them to produce their own hormones. Hormone action proceeds through obviously complex stages of interaction that are ultimately bound up with the entire regulatory mechanisms of the body.

Although it would be well beyond the aim of this book to review even briefly the large body of facts and concepts of the study of endocrinology, some other hormones should be mentioned for the sake of completeness. Not listed in Table 5–1 are such hormones as serotonin, released from the gastrointestinal mucosa, and regulating in part blood pressure. The small parathyroid glands of the neck region produce parathiomone, which regulates calcium and phosphate levels (which do affect nerve cells). The thyroid gland produces thyroxin, a hormone of very general and broad effects on metabolism through action in many places of the body: the thyroid gland is absolutely necessary for life. The pancreas produces insulin and glucagon which lower and raise blood sugar levels, respectively. To these

TABLE 5–1

Summary of Major Vertebrate Hormones and Hormone Groups Affecting Behavior (Known Primarily from Mammals)

ENDOCRINE GLAND	HORMONE OR HORMONE GROUP	PRINCIPAL PHYSIOLOGICAL EFFECTS*
Adenohypophysis (of pituitary)	gonadoptophins	st estrogenic H sec
	(a) Follicle-Stimulating H (FSH)	st ovarian follicle dev
	(b) Luteinizing H (LH)	st ovulation
		st estrogenic H sec
	(c) prolactin	st mammary dev
		st corpus luteum sec
	thyrotrophic H	st thyroid sec
	AdrenoCortico Trophic H (ACTH)	st adrenal cortex sec
	somatotrophin	
Neurohypophysis (of pituitary)	AntiDiuretic H (ADH)	reg blood pressure
		reg kidney
	oxytocin	st uterus dev
Cortex (of adrenal gland)	steroids	reg kidney
	(e.g., cortisone, aldosterone)	reg sodium balance
		reg carbohydrate metabolism
	other steroids	(see ovary, testis)
Medulla (of adrenal gland)	adrenalin	reg blood sugar, blood pressure
	noradrenalin	st sympathetic NS
Testis	androgen steroids	st testis dev
	(e.g., testosterone, androsterone)	
	other steroids	(see ovary, placenta, adrenal cortex
Ovary and placenta	progestins (from corpus luteum)	st gestation
	(e.g., progesterone)	
	estrogens (ovary and placenta)	st ovary dev
	(e.g., estradiol, estrone)	
	other steroids	(see testis, adrenal cortex)

* st = stimulates   dev = development
   sec = secretion   H = hormone
   reg = regulates

could be added hormones secreted from the kidney cortex, blood, damaged tissue, stomach, and upper intestine, all of which have important, if sometimes minor, physiological effects, to say nothing of the several "transmitter agents" of the nervous system, also properly called hormones, and so, important to proper nerve and brain function.

METHODS FOR STUDYING BEHAVIORAL EFFECTS

Studies dealing with hormonal effects on behavior have been largely empirically oriented, rather than theoretical; they are also fairly recent. Indeed, it was only in 1905 that the term hormone was coined by Starling.

Five lines of evidence have generally been marshalled in establishing the effects of hormones upon sexual, parental, and other behavioral responses: (1) first, and initially commonest, were studies establishing a correlation between the size of a gland and behavioral activity; for instance, gonadal weight and frequency of copulatory attempts. Breeding seasons received particular emphasis, for among seasonally breeding animals copulation may be limited to a relatively brief period during the course of the year. Red deer stags (Darling, 1937), for instance, are in rut only a few weeks out of 52. Only during these few weeks is there evidence of active spermatogenesis. F. H. A. Marshall in his classic book, *The Physiology of Reproduction* (1922), surveyed breeding seasons in animals from protozoans through vertebrates, summarizing what was known of seasonal, behavioral, and anatomical correlations.

Correlations, however, have limited predictive value, and experimenters soon wanted to induce behavior by manipulating hormonal mechanisms. (2) Thus, a second method for studying hormonal effects, that of injection of extracts from the glands, was used. (3) Closely allied to this was a third method, the induction of hormonal cycles by the injection of substances which stimulated the activity of the gland under study. (4) The reverse procedure was also tried: removal of the gland under study. This was a particularly convincing technique if followed by the re-implantation of the gland in question or re-injection of its hormone (*replacement therapy*). (5) Finally, the behavior of congenitally deficient animals, such as Riddle's (1935) "birds without gonads," were studied, nature having performed the surgery.

These five methods, although theoretically applicable to the study of any kind of behavior, actually found most application in sexual and parental behavior, as is shown by some examples below.

Sexual Behavior as a Case Study

Studies of reproductive behavior were initially concerned with the question, "Which hormones are involved in bringing about specific responses?" Studies involved fish, amphibians, reptiles, birds, lower mammals, and primates. Although some involved field work, the majority were performed on animals which could be kept easily in the laboratory. The studies were rigidly empirical, generalizations evolving only from a compilation of results, later summarized by reviews such as Beach's *Hormones and Behavior* (1948). Theoretical speculations were generally cautious.

Aristotle, as in most endeavors of the mind, appears to have been the first to record for posterity that gonads exert a strong influence on sexual activity: "the ovaries of sows are excised with the view of quenching their sexual appetites" (quoted by Beach, 1948: 30). Repetition of Aristotle's basic experiments on dozens of other species may seem a bit dull, but it has provided evidence that similar relationships hold throughout the vertebrates. Androgens such as testosterone secreted from interstitial cells in the testes increase the performance of sexual activities in males; similarly, ovarian hormones work upon female behavior.

Would that it were always so simple! It was noticed that as one ascended the phylogenetic scale from fish to primates, sexual behavior became much less directly dependent on gross blood levels of a single hormone. Indeed, sexual behavior of post-menopausal women is common in our society suggesting a fairly complete emancipation of sexual excitability from hormones (cited by Beach, 1951). Furthermore, the same experiments upon animals of different ages often yielded different results. Finally, other hormones than androgens and estrogens could be shown to have real, if small and sometimes complex, effects upon sexual behavior. For instance, adrenal activity seemed to promote sexual behavior and to do so mainly by its activating effects on general motor activities.

A rather special problem arises in sexual behavior in that two kinds of organisms within the same species execute different behavior patterns. Thus, it was necessary to test the obvious hypothesis that sexual differences were due to gonadal, and therefore hormonal, differences. Preliminary studies of hormonal effects on sexual behavior and sex differences had led Steinbach (1913) to propose *antagonistic actions of sex hormones*: androgens elicited male and suppressed female behavior, while estrogens had the opposite effects.

In general, the expectation "that when reversal of overt sexual responses occurs, the change is due to increase of the heterologous hormone" (Beach, 1948: 60) was confirmed. However, estrogen often elicited *male* behavior, and androgen, female responses; either hormone might produce both sexual reactions in the same individual. Thus, relatively simple views of hormonal action in sex, such as Steinbach's antagonistic hypothesis, could not cover certain facts. Part of the reason for the antagonism between gonadal hormones is that often only one kind of sexual behavior can be exhibited at once. One might say that the actual motor patterns are antagonistic, but this in itself offers no proof that the hormones actively suppress the behavior of one sex while activating that of the other.

HORMONAL-GENETIC INTERACTIONS IN SEX DETERMINATION

Systematic study of the effects of hormones upon sexual functions received its empirical impetus from C. O. Whitman at the turn of the century, although Whitman's major published works first appeared posthumously in 1919 (see also Chap. 2). Many of Whitman's published papers were collected and edited by O. Riddle, who has himself been a major contributor to this area.

Whitman recognized that the fertility of individual doves and pigeons varied with their age, the season, and different stages of the season. He also noted that the sex of the young from the first-laid egg of a clutch tended to be male, while the young from the second was female. Also, that the male represents a higher degree of differentiation (in gonadal embryogenesis, see below) than the female. His primary interest, however, lay in the basis of variation and related evolutionary phenomena. He believed he had evidence to demonstrate the existence of mechanisms for altering the germ plasm, thereby allowing the development of new heritable capacities. He did not focus upon the nature of these germ-plasm modifiers.

It must be remembered that the genetic and endocrine bases for sexuality had not been clearly established at the date Whitman wrote. Although Mendel did suggest that sex was established by hereditary factors, experimental proof was not available until the work of Doncaster in 1906 and that of Correns in 1907 (for a review of and references to this and the following early work in the genetics of sex, see Crew, 1953).

The notion that sex was dependent on the presence of an unpaired chromosome (the heterosome) was not fully developed until works by Wilson, Morril, Boveri, Julick, and Mubow appeared from 1909 through 1912. Almost at the same time, beginning with a series of papers published in 1911, Goldschmidt recognized the existence of intersexuality. By the early 1920's, when Witschi's work appeared, the humoral contribution to sexuality became evident. At least in birds and amphibia, the gonadal cortex and Müllerian ducts were seen to predominate in females, the gonadal medulla and Wolffian ducts in males. Since the medulla develops earlier than the cortex, it is possible for medullar products to suppress or retard the development of the cortex, and thereby also suppress the female characters associated with cortical secretions. The degree to which such hormonal suppression occurs is a measure of the strength of the male-determining factors of the chromosomes. An extreme example of this is as seen in freemartinism, a condition of female sterility in cattle. Freemartins result when bi-sexual twins are conceived. This is because twins develop anastomoses of the placental circulation. Since the male's gonads complete functional development ahead of the female's, the latter's system will be flooded by secretions from the male system. These secretions, in turn, will sufficiently block differentiation of the ovary as to lead to permanent sterility. A comparable situation in fowl, experimentally produced, of course, is known to produce an actual sex reversal (Crew, 1923).

PRENATAL EFFECTS OF PARENTAL HORMONES

Freemartinism and sexual development had been attributed to a difference in the timing of the production of sex steroids by the gonadal medulla and cortex. It was but a short step to envisage these hormones as directly influencing the higher centers of the brain. For instance, it was noted that if pregnant guinea pigs were injected with a male hormone, testosterone propionate, the female offspring would be permanently masculinized. Specifically, even the subsequent injection of estrogens would not cause the females born of the treated mother to exhibit lordosis (see Young, 1965; Gorski, 1971; Money and Ehrhardt, 1971 for reviews of the effects of perinatally administered hormones). Thus, the brain was seen as bipotential with regard to sexuality, but if the genotype was such as to allow androgen production prior to estrogen production, then the circulating androgens would alter or inhibit the activity of the "female" neural circuits. Since the time of maximum effectiveness of perinatally administered hormones varied from species to species (prenatal in guinea pigs, postnatal in rats) and was, in all cases, temporally restricted, it was further believed that the hormonal influence had to be excited at a particular "critical" period in development. The embryologic analogies should be obvious (if they are not, note Hamburgh, 1971). We will return to this point below and under Recent Developments.

In all, the mechanism of hormonal action in behavior was viewed as primarily increasing the sensitivity of a genetically organized behavioral or CNS pattern to external stimulation. Thus, Noble and Zitrin (1942) in their classical experiment of injecting baby chicks with testosterone observed nearly complete mating responses, including crowing, treading, etc. Hormones merely elicited the behavioral organization built into the organism.

A similar picture of the developmental interactions of hormones and behavior was indicated by Berg's (1944) study on the sex-specific urination postures in dogs. Females and young male dogs (under 19 weeks) squat, while adult males lift a hind leg. Injecting spayed females or puppies of either sex with androgens brought about the adult male posture, while castrated male puppies never switched to the adult method unless injected —and then they reverted to the female position after injections ceased.

Although other "neat" examples like this were uncovered, not all young animals responded to hormone treatment—or to castration and then hormone treatment as adults—in the same way. No developmental mechanism could be clearly specified. A typically enigmatic result is the incomplete cessation of sexual behavior after gonadectomy in adults, even though a castrated young animal fails completely to develop sexual behavior. The implication is that some interaction of hormonal action and behavioral development establishes a behavioral organization which then persists in the absence of hormonal influence.

SOCIAL-HORMONAL INTERACTIONS IN SEXUAL BEHAVIOR

That final sentence of the preceding paragraph must not be dismissed lightly. Consider this example: the level of sexual activity of a group of male guinea pigs or rats is individually measured. (Rats and guinea pigs do differ individually.) They are then castrated. Castration eventually leads to abolition of sexual responsiveness. Sexual behavior can be restored by the administration of testosterone propionate. However, the original individual differences remain, the hormone merely restoring the pre-operative level of activity. This simple result apparently depends on the animals having had some sexual experience prior to castration. At least, the ease with which sexual activity can be elicited by hormones varies according to whether there has been this pre-operative sexual activity. In cats, mating can continue for some time following castration in experienced toms, but inexperienced toms not only fail to mate but are also more refractory to androgen treatment (Rosenblatt, 1967; Ginsberg, 1965; Young, 1965).

Other evidence for the importance of experiential and social factors on the effects of hormones comes from the many studies of crowding, which demonstrate a consequent alteration in adrenal function with a concomitant influence on reproduction (Calhoun, 1963; Christian, 1963; Klopfer, 1973).

Much of the early work in this area was by O. Riddle, who is associated with the discovery of prolactin, which he and his associates announced in the 1930's (see Riddle, 1963, and Lehrman, 1963, for an interesting sidelight on this discovery.) Riddle's *magnum opus* (1935) was a detailed study of the relations between endocrines and physical constitution, a work which involved little attention to behavioral features. His prolactin studies, on the other hand, suggested prolactin was responsible for maternal behavior. The main point to consider here is that the administration of prolactin does not necessarily and invariably lead to broodiness (in pigeons or fowl) or other evidences of maternal behavior, nor does the appearance of maternal behavior necessarily indicate a heightened prolactin output. Both *priming* by other hormones and stimulation from the environment may mitigate, mimic, or alter the effects of prolactin. For example, implantation of progesterone pellets (into ring doves), according to Riddle, increases the anterior pituitary's output of prolactin and thereby induces broodiness. However, Patel's (1936) work (see also Lehrman, 1963; Hansen, 1971) shows that prolactin levels rise when a male merely views his mate incubating in an adjacent cage. Indeed, exteroceptive stimuli of a wide range may influence hormonal levels, including the sight, sound, or smell of newly born pups in rats (Mena and Grosvenor, 1971).

The role of hormones in the initiation of parental behavior is further complicated by the fact that males of many species (and females of some, e.g., the phalaropes) participate very little, if at all, in the care of the young. In some of these species appropriate hormonal treatment may induce parental behavior, just as prolactin induces broodiness in roosters of some races of fowl, or lactation in human males. In others, no amount of any hormones can promote the behavior usually manifested by the other sex.

In sum, earlier attempts to elicit sexual and parental behavior by manipulation of a particular "key" hormone have only been partially successful. A fuller picture of hormonal action demands more knowledge of the hormone's general effects, of the mechanisms whereby it acts, and of its developmental interactions in the growing organism.

CLIMATIC-HORMONAL INTERACTIONS IN SEXUAL BEHAVIOR

Beach (1948: 87) says that "the cyclic changes in the size of the avian gonad which correlate with the seasons of the year have been recognized since the days of Aristotle and Pliny." Aside from the fact that Aristotle believed that birds sought refuge beneath the sea in winter, it is still no surprise that Aristotle's contemporaries and subsequent men of similar intellectual bent shied away from an experimental investigation of migration such as that begun by William Rowan (1931).

Experimentation with migratory behavior is clearly a formidable technical problem. However, Rowan observed that the peak of gonadal recrudescence coincides with the peak of migration; gonadal hormones activated the migratory urge. The autumnal partial recrudescence of ovaries and testes in birds lent further weight to the hypothesis since the fall migratory departure might reasonably be expected to resemble spring return. (The

other possibility suggested by Rowan himself was that regression of reproductive activities in the fall released winter migration, analogous to the way
in which growth of hormone-induced activities in the spring released spring
migration.) The problem and hypothesis were formulated in Rowan's
famous little book (1931) *The Riddle of Migration.*

Rowan thereafter began an extensive program of manipulating hormonal flow in birds by lengthening the duration of the light period to which
captive birds were subjected each day. Although this ingenious technique
of mimicking one of the major environmental changes that accompany each
spring led to no clear-cut solution to the migratory problem, it did point
out very clearly how the environment can influence hormonal flow. As
might be imagined, the technique became quite useful in studying other,
especially sexual, behavior that is dependent on gonadal hormones.

To prove that a bird has been stimulated to migrate by manipulation
of its hormones outside the usual migratory period, Rowan released marked
birds in hopes of recapturing them. When juncos proved difficult to recap-
ture (the subsequent reporting rate for small land birds banded in North
America is rarely as high as one for 100 banded), Rowan turned to color-
dying larger birds. If lumberjacks in Canada reported seeing pink crows,
it comes as little surprise. Rowan's results certainly suggested that increasing photoperiod elicited northward migration (in the Northern hemisphere),
but, like other activities affected by hormones, no simple theory sufficed
as an explanation.

The gonadal hypothesis, with various modifications, was endorsed by
some (Bullough, 1945), but others who entered this fascinating field criticized the theory sharply (Wolfson, 1942, 1945). The criticisms were based
on subsequent work, several of the most important phases still under investigation, which showed that many factors influence migration. Increasing
day length, increasing ambient temperature, and other environmental
changes affect the animal through pituitary activity and perhaps the nervous
system directly. General physiological changes, including deposition of
internal and subcutaneous lipid reserves, precede migration. Weather *per
se* and perhaps the availability of celestial bodies for navigation en route
all affect migratory behavior.

The general physiological conditions bringing about *Zugunruhe* or
migratory restlessness that could be measured in a caged bird were specifically studied by Wolfson (1958). The pituitary is the "master gland" that
is stimulated in spring either by the direct effects of light or by the increased
activity of the bird during the longer days (or by both). The pituitary activity has multiple effects on the organism's metabolism, particularly in stimulating the deposition of lipids beneath the skin and surrounding internal
organs such as the gut. The correct migratory preparations themselves seem

to elicit the actual migration. In some species, sedentary populations could be shown never to develop the pre-migratory fat, while migratory populations of the same species did so. (For a review of field and laboratory studies on this question, see Helms and Drury, 1960.)

Perhaps fall migration is also brought about by similar physiological preparedness, but triggered by decrease in gonadal activity. This lack is, in turn, traced to lack of secretion of gonadotropins by the pituitary with increasingly shorter days. *Zugunruhe* can be delayed by exposing the bird to long periods of light in fall. Thus, it would appear that the physiological readiness that stimulates actual migration is brought about by any change in gonadal and/or pituitary activity.

MECHANISMS OF HORMONAL INTERACTION

To understand the action of a hormone, both the stimulation of its release and its effects upon behavior must be known. Neither was fully elucidated even for a single behavior pattern before 1950, but some likely generalities had been developed. Witschi (1935) distinguished physical and social stimuli that elicit hormonal flow in birds; F. H. A. Marshall (1942) later reviewed much of the evidence from mammals. Aside from Rowan's (1931) experiments on photoperiod, most of the evidence was circumstantial. Light, temperature, rainfall, food, and other inanimate factors, as well as the mate, companions, and offspring all seemed capable of eliciting hormonal effects—at least as judged by ultimate behavior. Later, A. J. Marshall (1954) began studying the effects of the behavior of one animal upon the secretions of its companion, the effects of the secretions upon the companion's behavior, and the ultimate feedback to the hormonal flow of the first animal. Christian (1963) and a host of others studied the changes in adrenal function that occurred as population grew and was subjected to greater social stress. Thus, a new field of hormonal research was born just after the end of the 1960's.

The other side of the coin, the mechanisms by which hormones affect behavior, is also complex. Beach (1948) separates four theoretical possibilities, and some evidence could be marshalled to support each. Hormones may influence behavior through their effects upon:

1. The whole organism (e.g., general activity level).
2. Morphologic structures employed in specific response patterns.
3. Peripheral receptor mechanisms.
4. Integrative functions of the central nervous system.

The last category is further specified: (1) control of the development of nervous organization, (2) control of periodic growth and regression of nervous elements, or (3) control of sensitivity to stimulation.

Hormones regulating many physiological parameters of the living animal can hardly fail to have some effect upon behavior through changes in bodily temperature, metabolic rate, or other mechanisms. Likewise, the hormonal effects upon genitalia or secondary sex characteristics (e.g., the soft-part colors of birds) are well-known. Changes in peripheral receptors and alteration of central nervous properties, in any of the forms suggested by Beach, must be considered as speculative, albeit quite possible. However, recent work involving the direct introduction of minute quantities of hormone into the central nervous centers has shown that specific acts may thereby be elicited, depending on the injection site and the material injected (Komisaruk, 1971).

Probably the most influential of the experiments that stressed the interrelatedness of the routes for hormonal elicitation and function were the studies by D. S. Lehrman (1965). He was able to show that (in ring doves) courtship, nest building, and incubation could all be induced by the injection of appropriate hormones in a particular sequence. However, the endogenous production of the same hormones could also be induced or facilitated by exteroceptive stimulation. A similar complementary and feedback relationship between exteroception, particularly involving the social milieu and hormone production, was described in canaries by Hinde (1970). Further, a hormone could both alter peripheral sensitivities and central function. A hormone whose release is induced by the sight of eggs in a nest will then stimulate the development of a brood patch. This defeathered, vascularized patch of skin has its tactile sensitivity altered so that the contact with the smooth egg shell becomes "pleasurable," i.e., the valence of inputs from incubation differs from that obtaining when the peripheral receptors are either in a different state or narcotized. Bird-sitting-on-eggs, in turn, provides a new exteroceptive stimulus for the mate, which represents yet another source of feedback control. Finally, some hormones are known to have specific effects upon particular regions of the brain as well as peripheral influence. Direct application of minute quantities provokes particular behavior patterns. It is presumed that if a hormone is circulating in the blood, it will be able to reach these centers. The close affinity between hormones and neurons needs to be underscored. Both neural and hormonal tissues are derived (in major part) from neural crest ectoderm. The nerve cells themselves are functionally indistinguishable from endocrine cells. What are the neurotransmitters if not hormones, though their quantity and milieu are such as to preclude their moving much farther than the distance of the synaptic cleft? The fact that there may be a duplicity in behavioral control mechanisms must be less than astounding (more on this subject appears in Chaps. 11 and 12).

Summary of Conclusions

BEACH'S "GUIDING GENERALIZATIONS"

No better conclusion to the pre-1950 results on hormonal research exists than Beach's (1948: 250–255) "guiding generalizations":

1. None of the behavioral responses thus far investigated have been found to depend on one and only one hormone.
2. No hormone is yet known which produces one and only one effect upon the organism.
3. Hormonal control and a complete behavior pattern appear to be mediated by several intervening mechanisms rather than a single critical one.
4. Different behavior patterns may vary in their degree of dependence on endocrine products.
5. Every behavioral response constitutes a reaction of effector mechanisms to internal and external stimuli; the effector mechanisms are neuromuscular or neurohumoral units whose functional characteristics are determined jointly by genetic constitution, by previous excitation, and by the chemical nature of the internal medium.

With these conclusions we may contrast the "newer" look at hormonal-behavioral relations that began to emerge shortly before 1950.

THE NEW LOOK IN HORMONAL RESEARCH

In 1915, Walter Cannon, the famous Harvard physiologist, published an important book dealing with the physiological bases of "emotional" behavior (*Bodily Changes in Pain, Hunger, Fear, and Rage*). Although much of this work would be today considered more in the sphere of physiology than ethology, his studies of hunger, thirst, blood circulation, and other bodily functions presented an important new principle—*homeostasis.* Cannon's principle states that all parts of the controlling physiological system are interrelated and self-regulating.

The discovery of gonadotropins secreted by the pituitary in response to environmental stimulation mediated through the hypothalamus was particularly important to the understanding of hormonal feedback relationships. For instance, it was found that the gonadotropins of the male vertebrate activate the interstitial cells of the testes to produce androgens, which in turn stimulate spermatogenesis. As the androgen blood level rises, this

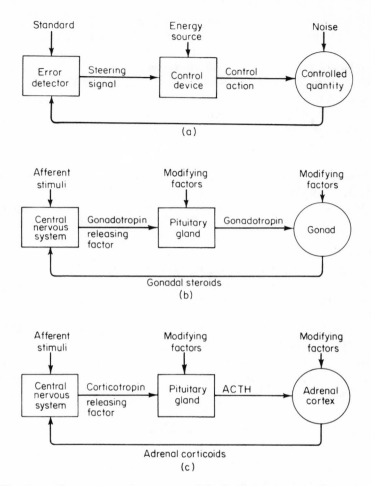

Fig. 5–1. Comparison of concepts of feedback systems in physics and neuroendocrinology. (a) Feedback system as of a thermostat controlling a furnace; (b) analogous hormonal system controlling gonadal functions; (c) corresponding control system of the adrenal cortex. (After Scharrer and Scharrer, 1954.)

hormone "feeds back" to the pituitary to suppress secretion of gonadotropin. Thus, once spermatogenesis is complete (and mating has occurred), the male is again ready to begin a new reproductive cycle in response to environmental stimuli. The female mammal's cycle is even more complex, involving several steps of feedback interrelationships. The analogy of neuroendocrinological systems with physical feedback systems can be seen in diagrams (Fig. 5–1).

Toward the end of our survey period the notion of feedback had been extended to behavioral-hormonal relations in the preceding and related instances by A. J. Marshall in his study of bowerbirds (e.g., his 1954 book, *Bowerbirds: Their Displays and Breeding Cycles*). Thus, the display of a male could initiate hormonal secretion in a female, the secretion affecting her behavior as well as internal organs. Her changed behavior could then feed back to the male, inducing in him hormone changes which alter his own behavior. Much of the literature (published before 1950) relevant to behavioral-hormonal interrelationships was reviewed by Lehrman (1959; 1965).

The other part of the new look in hormone research concerned the obliteration of the firm distinction between neural and hormonal regulation of bodily functions. The Scharrers (e.g., in their 1954 review) showed that many neurons produced hormonelike chemicals. In fact, all synaptic transmitter substances could be considered hormones. Furthermore, the adrenal medulla is embryologically of neural origin and secretes the same *neurohumor* as the sympathetic postganglionic nerve cells. Thus, some endocrine glands may be looked upon as a concentrated terminal part of a neuron whose neurohumor is secreted into the blood to be carried to many places at once. Conversely, nerves can be considered elongate endocrine glands whose function is to deliver small amounts of hormones to a specific site. Many of the *trophic hormones* of the anterior pituitary, including gonadotropin mentioned above, appear to originate in the hypothalamus and migrate to the pituitary where they are temporarily stored for later release.

Some Recent Developments

A major reorientation in studies relating hormones and behavior has taken place so recently (the 1960's) that current workers will likely require the next decade to test the ideas already announced.

Work continues on the notion that hormones present prenatally or early in life may alter brain organization. Although there is much to support this view (Lisk, 1967; Gorski, 1971), the specifics of the effect are as yet undescribed. Nor has everyone been convinced. Frank Beach, the dean of the field of hormones and behavior, has remained skeptical (Beach, 1971).

Identification of the hormones involved in particular behavior pathways and their exteroceptive control continues to occupy workers. Terkel and Rosenblatt (1972) cross-fused the blood supplies of freely moving rats to demonstrate the time of appearance of blood-borne "maternal" factors. Virgin rats receiving blood from parturient females would become "maternal." Their own blood, however, was not capable of inducing maternal behavior in other rats, demonstrating again the dual or complementary involvement of humoral and nonhumoral factors.

Finally, the range of relevant exteroceptive stimuli has been broadened to include social signals. The importance of these had been clearly implied by the work of Lehrman (1965) and others, already cited. However, Michael and Keverne (1968, 1970) were able to relate social signals more explicitly to hormonal responses. They isolated a secretion from the vaginal tissue of estrus rhesus monkey females which elicited sexual behavior in males. In short, the animals are aware of one another's sexual status, underscoring the role of hormones as synchronizers of social activity (including sexual behavior!) and their interrelatedness with exteroceptive inputs in the larger sense. Of course, the study of pheromones, to which this work relates, is also continuing, particularly among insects, and also in connection with the phenomena of pregnancy blocks or facilitation in mammals (review by Parkes and Bruce, 1961).

Overview

Early research on hormonal effects upon behavior was directed to the question, "Which hormone brings about what behavior?" Broadly comparative experiments showed that gonadal hormones tended to elicit sexual behavior (androgens, male behavior; and estrogens, female behavior) and pituitary lactogenic and other hormones tended to elicit parental behavior. But hormonal-behavioral correlations were rarely perfect, becoming less so in mammals than in lower vertebrate forms. Hormonal action was found to be even more complex in migratory, aggressive, and other nonreproductive behavior. Nearly all behavior is influenced in some way by hormones, and nearly all behavior in some way influences hormones.

Hormone flow is elicited by a very broad range of physical and social external stimuli. Hormones may act directly on central neural mechanisms underlying behavior, but they also certainly act on particular organs employed in behavior, systematically, and possibly on peripheral sensory receptors. Although hormones often act to elicit a pre-organized behavioral response through increasing its susceptibility to external stimulation, hormones appear to act also in more complex ways during development of behavior.

Recent research has centered on the feedback and synergistic relations of hormones and the interrelationships of hormone secretion and external stimuli. The study of the ways in which hormones affect behavioral development through effects on peripheral organs as well as on differentiating neural tissue will require a vigorous new approach.

Selected Readings

Beach, F. A. 1948. *Hormones and Behavior.* P. B. Hoeber, New York. 368 pp. (The classical source for ethologists on effects of hormones on behavior of animals, especially vertebrates.)

Beach, F. A. (ed.). 1965. *Sex and Behavior.* Wiley, New York. (A compilation of many of the important recent studies.)

Hinde, R. A. 1970. *Animal Behaviour: A Synthesis of Ethology and Comparative Psychology,* 2nd ed. McGraw-Hill, New York. (Includes an excellent summary of Hinde's own work on the hormone-behavior problem.)

Rowan, W. S. 1931. *The Riddle of Migration.* Williams and Wilkins, Baltimore. 151 pp. (The famous account that can be said to have begun serious studies of hormonal and physiological aspects of migratory behavior.)

Wheeler, R. E. (ed.). 1967. *Hormones and Behavior.* Van Nostrand, New York. (This collection includes many of the older artitcles cited in this chapter.)

Young, W. C. (ed.). 1961. *Sex and Internal Secretions,* 3rd ed. Williams and Wilkins, Baltimore. (An impressive tome by many expert contributors. D. S. Lehrman's chapter (pp. 1268–1382) is of particular interest to ethologists, for it compares birds and mammals and further relates these studies to problems of learning, genetics, and evolution. The 1939 edition also is a valuable reference for hormonal studies of sexual behavior.)

Chapter 6

UMWELT UND INNENWELT DER TIERE

Studies in Perception and Sensory Mechanisms

The "Umwelt" of von Uexküll

If you look up *Umwelt* in a dictionary, you will find a translation such as "environment" or "the world around us." This is *not* what von Uexküll meant. Rather, the *Umwelt* is the world around us *as we perceive it*. Each animal has its own *Umwelt*, characteristics of which are dictated to a large degree by the animal's sensory mechanisms. It is this link between perception and its sensory basis which forms the unifying theme of this chapter.

This chapter really deals with a rather limited portion of Jacob von Uexküll's total conception of behavior, as put forth in his classic *Umwelt und Innenwelt der Tiere* (1909). This book has a strong vitalist outlook but is no less valuable for that. Among the important points von Uexküll raises is the concept of behavior as a feedback relation with the environment. A perceptual sign or stimulus (*Merkzeichen*) of the environment is received by the animal's receptor (*Merkorgan*); the internal information (*Wirkzeichen*) is transferred to the effector (*Wirkorgan*). The action of the effector provides an "effector clue" (*Wirkmal*) that extinguishes the "receptor clue" (*Merkmal*) and thereby terminates the behavior (cf. the reafferance principle, Chap. 4).

Von Uexküll tries to translate the *Umwelt* of simple animals, such as the paramecium and sea urchin, into human perceptual terms and then attempts to describe perceptual worlds of animals with more complex sensory systems. His analogical conception can be shown by a simple example. If an animal is color blind but otherwise has eyes similar to our own, the animal's *Umwelt* can be approximated by a black-and-white photograph of its physical environment.

However, the *Umwelt* can be determined by factors more complex than mere receptor structure, as shown by von Uexküll's example of the perception of an oak tree by different species. To the forester the tree is an inanimate object to be measured; to a small girl it is an object to be feared, for she sees in the gnarled bark an evil face; to the fox the tree is a comfortable home in the roots; and to the owl it is a roosting place in the branches. Thus, the *Umwelt* of these animals all having similar eyes depends in part on spatial aspects of the tree to which they attend and on their past experience.

Aside from studies by his students (e.g., Lissman, 1932), von Uexküll's book has had relatively little effect on animal behavior studies compared with the great originality of its content. Part of the reason is that von Uexküll's analogical concepts met with the same difficulties as studies of the "mind" (see Chap. 3). If we are to infer the animal's *Umwelt* from the structure of its receptors and its behavioral responses to stimuli, then why not just study receptor structure and function and its relation to behavior *per se* without unnecessary inferences? The remainder of this chapter is devoted to such operational studies. Few received direct impetus from von Uexküll, but they may be considered to be studies in the same tradition.

Perception in Bees and Bats

As examples of the intimate relation between perception and behavior consider two famous studies, one on bees and the other on bats. Such studies on animals with specialized sensory capacities facilitate learning about sensory systems in general.

THE VISION AND LANGUAGE OF BEES

Bees can be readily trained to take sugar water from a dish. Von Frisch (1927, 1955) set beneath such a dish a colored (e.g., blue) card so that the bees became conditioned to finding food at such a color. The critical test came when he removed the food and placed the blue card in an array of gray cards, ranging in shade from black to white. If the bee has no color vision, but has learned the blue card only on the basis of its brightness, it should then confuse the blue card with one of the intermediate grays. Bees are not confused, however. Wherever in the array the blue card was placed, the returning bees landed on it alone in search of food.

Von Frisch's (1923) second major discovery about bees is one that transcends mere perception but is yet intimately related: bees emerging from the hive find a feeding station with incredible accuracy. Although the first to arrive at the station may appear to find it by chance, in a few minutes new bees can be observed to fly a straight line to the station. Somehow the scouts must communicate their find to their fellows within the hive. It is of advantage to have such a communication system, for the natural food source of bees is the nectar of flowers. The location of ripe flowers changes from day to day, even from hour to hour, since new flowers open up and others close as the day progresses.

Von Frisch broke the communication code of the bees by watching returning scouts in a glass-covered observation hive. The returning bee *dances* in a circle and then reverses and makes a circle in the other direction. The liveliness of the dance indicates the richness of the food source. Other bees, clinging to the scent of nectar on the scout, go out in search of the same scent.

If the distance to the food source is great (more than 50–100 meters), the scout's dance becomes a figure eight (*Schwänzeltanz*) with the bee running in a straight line (*Schwänzellauf*) between the loops. The frequency of complete *Schwänzellauf* performances indicates the distance to the food. When the food source is about 200 meters distant, waggling dances occur at the rate of nearly 40 times per minute, while at very great distances (e.g., 6000 meters), the rate is only 4 times per minute. (The actual relation between the dance frequency and the distance varies from one species or race of bee to another.)

A more recent study by Wenner (1962, 1967) contends that since the hives are dark inside, the bees cannot see the figure-eight dance. Instead, Wenner proposes that the bees respond to the sounds emitted during the "straight run" (*Schwänzellauf*). As the dancer performs his dance, the "recruit" bees pick up sound vibrations by placing their antennae on the thorax of the dancer. However, as Wenner points out, the thorax does not trace a figure-eight configuration. He illustrated this by photographing a dancer with a white spot painted on his thorax. This convinced Wenner that the visual effects of the figure-eight dance do not provide the bee with enough information on the location of the food. Wenner has shown through sound spectrograph analysis that the average number of pulses and the average length of the sound trains during the particular dance are proportional to the distance to the food source. Further investigation led Wenner to the conclusion that bees do not respond to sounds transmitted through the air but must receive the vibrations through their feet and/or antennae. This enables the recruits to follow the orientation and vibrations of *their* dancer since other dancers may be performing only inches away. This work has not seemed convincing to von Frisch, and the issues it raises are still a matter of controversy.

The *Schwänzellauf* also indicates the direction to the source. Bees normally dance on vertically placed hives. If the food source is in the direction of the sun, the *Schwänzellauf* points directly upward; that is, the figure eight lies on its side, and the bee moves upward during the straight portion between the loops. If the food source is in a direction 90 degrees to the right of the sun (for instance, the food might be to the north in the evening), then the returning bee dances with the *Schwänzellauf* pointed 90 degrees to the right on the vertical hive. And if the food is in the opposite direction from that of the sun, the bee dances downward. Thus, the insect's nervous system transforms—by a process still not understood—the direction of food with relation to the sun to the direction of the *Schwänzellauf* with relation to gravity.

Many of von Frisch's other discoveries are related to this dancing. For instance, it can be shown that bees have a time sense because if they continue to dance for a long time, the direction of their *Schwänzellauf* changes to compensate for the movement of the sun. Some other discoveries concerning vision stem from the fact that bees orient the *Schwänzellauf* directly to the food source when the hive is placed horizontally and light is allowed into it.

Von Frisch also proved that bees see a "color" that we do not: ultraviolet, or the wavelengths of electromagnetic spectrum just shorter than blue and violet wavelengths. By placing filters over the horizontal hive, von Frisch found that filters that allowed only ultraviolet (UV) and no "visible" light to enter caused no disruption of the orientation of the *Schwänzellauf*, but filters excluding both UV and visible wavelengths caused immediate distortion. Thus, bees can distinguish between two kinds of flowers that each look white to our eyes: those that reflect UV in addition to our visible spectrum. It can also be shown by artificial training methods, such as those mentioned above, that bees can distinguish these two "whites."

Finally, a major discovery of von Frisch's was that bees can perceive polarized light; it had been previously thought that no animal could do so. Light waves oscillate at right angles to the direction of their travel. Polarizing filters exclude light oscillating in all but a single plane, and the transmitted light appears dimmer to the human eye. We cannot distinguish which plane of oscillation the filter is transmitting because two filters placed side by side and oriented so that they transmit perpendicular planes of light look equally dim. (If they are juxtaposed, they exclude all light, of course!) But bees can distinguish the orientations.

Von Frisch noticed that bees could properly orient their *Schwänzellauf* on the horizontal hive if only a patch of blue sky, and not the sun, were visible. Light from the sky is polarized in various patterns in relation to the position of the sun because of filtering effects of the atmosphere. Bees "know" the polarizing pattern and can "calculate" from it the position of the sun and then orient their dances accordingly. This fact von Frisch proved by placing a polarizing filter in various orientations over the hive and thereby systematically altering the direction of the *Schwänzellauf*.

The studies of sound communication in bees have not been limited to the orientation dance. Wenner (1962, 1967) also examined "piping" of workers and the "toots" and "quacks" of queen bees. If a hive is briefly disturbed, the workers emit short beeps at about 500 Hz, which apparently calms down the hive. Wenner confirmed this by playing a recording of this piping to a group of disturbed bees. Young, developing queens produce a "quack" of lower frequency than the queen's "toot." These virgin queens are held back in their cells by the workers and are released one at a time. They fight with the old queen bee, who usually kills them. Wenner played recordings of these sounds to various queen bees. His findings confirmed that the "tooting" announces the presence of an established queen in the hive, and "quacking" reports the presence of challengers ready to be freed from their cells. This information guides the worker bees in their responses to the virgins.

Many of the phenomena discovered through the keen observation of bees living naturally and subsequent simple experimentation have recently been uncovered in other arthropods. To cite such later studies here would be unnecessary, but they argue that widely occurring behavioral phenomena are still awaiting discovery by critical minds and keen eyes.

THE ECHOES OF BATS

During the winter of 1938, a Harvard undergraduate named Donald Griffin approached the famous physics professor G. W. Pierce with an unusual request: he wanted to find out if bats produced sounds that the human ear did not hear. Pierce had the necessary equipment to record sound of very high frequencies (now called *ultrasonic frequencies*). Griffin's initial discovery that bats did produce such sounds and their further elucidation in collaboration with another student, Robert Galambos, wrote *finis* to several centuries of speculation. They proved what Griffin (1958) rightly, albeit modestly, attributes to the eighteenth-century biological genius Spallanzani: bats orient by auditory means. The history and sidelights of this fascinating story are reviewed, along with the experimental work of Griffin and his students, in Griffin's classical *Listening in the Dark* (1958).

By taking bats into darkened rooms filled with vertically strung wires, Griffin was able to show that the animals avoid striking the wires in flight. Many experiments have shown that it is the *utrasonic echoes* from these wires that facilitate avoidance. The bat literally hears its way in the dark. In anechoic chambers the bat becomes disoriented, as is the case when its ears are plugged.

Griffin thinks of himself as an "experimental naturalist" and has taken the lead in developing a field of sensory biophysics that involves measuring biological phenomena with the tools of physics. Imagine his station wagon parked near a place known to be frequented by bats; on the tailgate sits a cathode-ray oscilloscope powered with a gasoline generator (connected to a car muffler to dampen noise) and photographed with a portable motion picture camera. At the end of a long cord stands Griffin with a microphone in a parabolic horn, pointing it at bats swooping by in the dark in order to record their sounds.

Such studies produced evidence for echo-location avoidance of natural objects in the field and for the use of ultrasounds in locating the flying insects upon which bats feed. A further modification was necessary to determine whether a fish-catching bat of Central America also uses echoes to spot its prey. For this, Griffin took his microphone into a dugout canoe with a long "umbilical cord" to the generator and oscilloscope on the shore. Present evidence indicates that the fish-eating bat hears echoes only from parts of the fish protruding from the surface (e.g., dorsal fins). The original pulse and then the returning echo lose too much of their energy at the surface-air interface for underwater parts of the fish to be detectable.

The comparative studies carried on by Griffin and his students such as A. Novick and A. P. Grinnell, as well as by workers in Europe (Dijkgraf, Möhres), have elucidated the nature of ultrasonic pulses in a huge number of bat species with various ways of life (for further details, see Novick, 1971). It turns out that whirligig beetles, whales, and even some birds also find their way by means of echo-location. Griffin's studies have had great application to problems faced by blind persons and detection of objects by instruments such as radar and sonar. The discovery of ultrasonic echoes reminds us again of von Uexküll's dictum that the perceptual world of animals is not necessarily that of our own!

Sensory "Windows" and Behavior

Other studies of the perceptual basis of behavior must be mentioned to show the fruitfulness of this approach. Many facts and discoveries of sensory physiology are, of course, relevant to behavior, but until recently few have had any direct relationship to specific behavior patterns (see Recent Developments for this chapter). Some examples relating to the various "sensory windows" (Griffin's expression) follow.

VISION AND AUDITION

The light intensities at which owls could find immobile prey were measured by Dice (1945). The owl can see in very dim light, but it apparently cannot perceive infrared rays, as is sometimes suggested. (Recently, R. Payne has demonstrated that the barn owl can catch prey in complete darkness, using only its acute sense of hearing; Payne and Drury, 1959 and Payne, personal communication.) Dice's study is another dimension of studies of animal sensory mechanisms underlying behavior; bees see wavelengths we cannot, and owls see in light so dim that we would be blind.

The ranges of sensory capacities are not the only parameters of perceptual processes, however. Wolf and Zehrren-Wolf (1937) showed that bees perceive flowers, not by the shape of the blooms *per se*, but by the flickering effects that the serrated edges create on the compound eye of the moving insect. The Wolfs continued their investigation of *flicker* in collaboration with Crozier at Harvard, and the team produced many papers—unfortunately never really summarized in one authoritative review—that appeared chiefly in the *Journal of General Physiology* during the 1930's. Among their discoveries was that the shape of the eye of insects affected their perception of flicker (artificially produced through the movement of alternating stripes of black and white on a rotating drum). Their studies had much greater immediate relevance to strictly physiological problems, but the body of facts they amassed is useful in analyzing behavior patterns elicited by flicker.

Hearing is the other sensory window to which man most consciously attends, and similar kinds of studies have been done in this modality. Karl von Frisch, for instance, proved that fish can hear by training blinded individuals to come for food in response to sounds produced in air by tuning forks and whistles. Destruction of the parts of the middle ear associated with hearing in other vertebrates abolished the responses, while damage to portions of the middle ear having to do with the maintenance of equilibrium had no direct effects.

On a slightly different tack, Marler (e.g., 1955) found that the communicative sounds produced by birds could be correlated with the ability to localize sound. Sound source localization depends on comparing several parameters of the sound between the two ears, including differences in phase, intensity, and time of arrival. Intensity differences are most useful at high frequencies, phase differences at low frequencies, time differences at all frequencies. Marler has shown that mating calls (whose source should be easily located) differ radically from alarm calls (which are best not traced!); the former are generally composed of many frequencies and are broken up into short phrases, frequently repeated. Alarm calls, on the other hand, are generally high, thin notes, beginning and ending imperceptibly.

OTHER SENSES

Some fish produce an electric field through organs in the tail. The various species of electric fish apparently investigate their environment and communicate with their fellows in a manner somewhat analogous to the bat's sonar, by sensing the disturbances in their electric field (Lissman, 1958; Black-Cleworth, 1970). Another strange sensory window has been elucidated through the work of Noble and Schmidt (1937). Pit vipers sense with their pit organs the presence of small mammals by perceiving the infrared radiation produced in metabolism. Thus, snakes can "see" infrared, as bees do ultraviolet.

The chemical senses have been not ignored. Dethier (1955) and Hodgson (1961) have worked out the physiology of chemoreceptors in the legs of blowflies, trying even to decipher how the sensory information is processed as it moves centrally toward the cephalic centers. Wilson (e.g., 1962) has given an intriguing account of how chemical trails facilitate communication about food location in ants. On the basis of their behavioral use, Wilson (1962) predicted the physical characteristics that certain chemicals ought to have; for instance, a warning substance must be rapidly diffusing, but a trail substance must diffuse slowly. Through biochemical analysis of the actual substances used, he confirmed these predicted physical properties. Similar analyses of the exudates from particular glands have allowed Müller-Schwarze (1969) to analyze olfactory communication in deer and antelope. The importance of chemical signals in intraspecific communication has been acknowledged by the coining of a new term, *pheromone*, which stands for olfactants with a communicative function. The importance of pheromones in regulating behavior in primates is indicated by studies of Michael and Keverne, 1968 (more on this topic under Recent Developments, below).

This discussion would be incomplete without a mention of how the internal environment of an animal can affect its responsiveness to stimuli. C. P. Richter (1945) showed that rats deprived of certain dietary requirements (e.g., a salt) will make up the deficit by selecting food containing the deficient substance from among foods lacking it or possessing it in lower concentrations. The sense of taste is probably of immense importance in the behavior of complex organisms, but it has received unfortunately little ethological investigation (but see Garcia, 1973).

Finally, von Holst's (1950) experiments on equating the effects of two different sensory inputs point the way to a "second order" of sensory investigation. Angel fish swim right side up by (1) "putting gravity below them" and (2) "putting the sky above them." Von Holst could vary the gravitational and visual inputs separately by (1) putting a centrifugal force on the fish by means of a whirling chamber with running water and (2) shining a light into the aquarium from any angle (in a darkened room, of course). Thus, he was able to show how much one sense, relative to the other, was responsible for normal posture; he could even make a fish swim upside down! Furthermore, he could alter the relative importance of the two senses by adding food extract to the water. The angel fish hunts by sight and relies more on the light input during feeding than at other times.

Many other facts of sensory physiology uncovered during the first half of this century are, of course, relevant to animal behavior. As the final section will demonstrate, the gulf between sensory function and behavior is closing. As a physiologist (H. B. Barlow, 1961: 233) has put it: "It is foolish to investigate sensory mechanisms blindly—one must also look at the ways in which animals make use of their senses."

PATTERN PERCEPTION

That the gap between sensory physiology and behavior is closing can be shown by some examples of studies reported since 1950. First, in vision Goldsmith (e.g., 1960) has succeeded in tying the color vision of bees to specific visual pigments and receptor physiology. In an exciting study, Maturana, Lettvin, Pitts, and McCullough (e.g., Maturana et al., 1960) have found five kinds of ganglion cells (third order neurons) in the retina of the frog's eye. Four of these are independent of ambient light levels and respond only to certain stimuli of specific shape and movements. For instance, one kind of ganglion cell responds only to a moving curved boundary whose convex side is darker than the concave side; such peripheral filtering may "code" the stimulus situation produced by an insect (upon which frogs feed). Recently the team has discovered two types of neurons in the visual system of the frog's brain (in the colliculus), one of which responds to "novel" stimuli and the other of which responds to "interesting" stimuli (e.g., insects) and continues to respond as long as the object is in the visual field. Similar organization of the visual sense has been discovered in the pigeon by Maturana and Frenk (1963) and in the cat by Hubel and Wiesel (1959; see also Chap. 11). What these studies have in common and what differentiates them from studies that bear principally on an animal's sensitivity to particular sounds, scents, or wavelengths is their emphasis on the sense organs as organizing systems. For instance, Muntz (1964) in his study begins with a criticism of the notion that the retina is analogous to photographic film, with each receptor cell acting as a grain of the emulsion. Instead, he proposes that the retina is more like a filter that sends only the most useful information to the brain. In order to determine what kind of information is important to a frog, one must approach the problem from the frog's point of view. This accords entirely with the views expressed by most of the leading students of sensory function. Thus, the fact that frogs can distinguish blue from other colors and preferentially jump toward blue lights is related both to the existence of "blue-responding" fibers projecting to the midbrain and the utility of a leap toward water when danger threatens.

In hearing, too, recent work is connecting behavior and physiology. Grinnell (1963) has demonstrated the physiological basis of bat hearing uncovered by Griffin. Recording evoked potentials from the bat's brain, Grinnell has found mechanisms for frequency discrimination, loudness levels, spatial location, and even some for recognition of the bat's returning echo from among thousands of other similar sounds produced by its companions. Along parallel lines, Roeder and Treat (e.g., 1959) and Roeder (1971) have shown that the three-celled tympanic organ of noctuid moths is "tuned" to receive the echo-locating pulses produced by the bats that prey on the moths. Behavioral experiments have revealed a number of escape ploys used by moths when hearing artificial bat sounds, while field experiments recording from both tympanic organs simultaneously have elucidated the physiological correlates of recognition and localization of the bat cries.

Some Recent Developments

Three areas of recent activity pertain to ultrasounds, smells, and magnetism, all fields in which one can fairly say that progress depended on prior technological advances. The development of high-speed, narrow gap portable tape recorders and high-range microphones was a necessary antecedent to the spate of studies of ultrasonic communication and the confirmation of its role in maintaining contact between mother and young (e.g., Sewell, 1970). The importance of ultrasounds had, in fact, been postulated much earlier (Zippelius and Schleidt, 1956; Calhoun, 1963), but only now are these earlier notions being adequately tested.

For the sense of smell, gas microchromatography has been the analogue of the tape recorder. Even for workers who do not particularly care what the chemical nature of their olfactants or pheromones is, the possibility of establishing a chemical identity seems to have been a stimulus. In any event, the use of scents in marking territories, individuals, eliciting or dampening sexual behavior, and other types of social behavior has become a suddenly popular and important area for research. The phylogenetic antiquity of the olfactory areas of the brain has led many to suspect olfactory inputs to be of especial, if nonobvious, significance in human likes and dislikes. (Of course, olfactory responses in insects have long been a focus for agricultural workers interested in insect control. The use of pheromones or sexual attractants as baits in traps is of major interest to entomologists.)

The renewed interest in magnetic sensitivities corresponds to the discovery that homing pigeons and some songbirds may indeed utilize positional information provided by the earth's geomagnetic field (see Chap. 8, and reviewed by Freedman, 1973). This is certain to be a subject of intense inquiry for the next several years, for few clues yet exist as to the nature and sensitivity of the putative "sense organs for magnetism."

The other area of expanding interest is less easily explained in regard either to technical advances or to the insights that have come from advances in perceptual studies. It pertains to the perception and use of space. Perhaps the rise in interest corresponds to the accelerating loss of space that results from ever-expanding populations!

In any event, the issue of space perception, once largely the domain of psychologists interested in visual systems, is now being considered by a wide range of biologists and sociologists. Some of the latter (Sommers, 1969) are concerned with individual behavioral effects of changes in spatial orientation and perception; others (e.g., Aronson and Rosenbloom, 1971) with the development of space perception in infants; still others with the relation between spatial organization of societies and the character of the substrate (Klopfer, 1973). The human implications of these studies will doubtless be actively and thoughtfully reviewed by ethologist and humanist alike (e.g., McInnis, 1969).

Overview

Von Frisch discovered that the visual world of bees is indeed different from our own. The bee has color vision that includes ultraviolet, a color outside our visible spectrum; the bee converts the direction of a food source in relation to the sun to the direction of *dancing* with relation to the inverse of gravity, thereby communicating food discoveries to its conspecifics in the hive; the bee has a time sense, and the bee perceives the patterns of polarized light in the sky.

Griffin discovered that the world of "blind" bats is perceived by echoes of their extremely high-frequency sounds outside our own hearing range. Similar studies of all the "sensory windows" show how an animal's perceptual world dictates the behavioral response to its environment. Von Uexküll's concept of the *Umwelt* (the world as the animal perceives it) is finally commanding the experimental attention it has for too long been denied.

Selected Readings

Griffin, D. R. 1958. *Listening in the Dark*. Yale Univ. Press. (An exciting account of the author's studies of echo-location and related problems in historical perspective. A shorter, less technical account is found in Griffin's *Echoes of Bats and Men*, published in 1959 by Doubleday-Anchor.)

von Frisch, K. 1927. *Aus dem Leben der Bienen*. Springer-Verlag. Berlin. (An intriguing book that brings together natural history and experiments on bees. The revised fifth edition of 1953 was translated into

English by D. Ilse and published in 1955 under the title *The Dancing Bees* by Harcourt, Brace and Jovanovich.)

von Uexküll, J. 1909. *Umwelt und Innenwelt der Tiere*. Springer-Verlag, Berlin. (A second edition of this classic was published in 1923, and a shorter, revised version in 1934 under the title *Streifzüger durch die Umwelten von Tieren und Menchen*. The latter work was translated into English by Claire Schiller for *Instinctive Behavior: The Development of a Modern Concept*, pp. 5–80.)

The most recent reviews of olfaction in mammals are:

Cheal, M. L., and R. L. Sprott. 1971. Social Olfaction: A Review of the Role of Olfaction in a Variety of Animal Behaviors. *Psychological Reports* (Monograph Suppl.), **29**: 195–243.

Ralls, K. 1971. "Mammalian Scent Marking," *Science*, **171**: 443–449.

Chapter 7

THE
SOCIAL LIFE
OF
ANIMALS

The Genesis and Development of Social Behavior

Introduction

Certain of Darwin's arguments concerning organic evolution rested upon the assumption that there was competition between organisms for a limited supply of resources such as food or nesting sites. Competition, of course, is almost never expressed directly through combat, but proceeds indirectly. Its outcome is seen as differences in proportionate contributions to the gene pool of future generations (cf. R. A. Fisher, 1930). A more anthropomorphic view of competition, however, proved generally pleasing during the mid-nineteenth century. Thus was born the un-Darwinian concept of "nature red in tooth and claw," a concept which implied direct aggressive competition and was scarcely at all what Darwin described. Nonetheless, as Darwin's theories of natural selection and evolution gained ascendency, Darwinism was increasingly interpreted as providing a natural sanction for current social and economic doctrines. "Survival of the fittest" became an excuse for the maintenance of ruthless social practices. This in turn led to an increasing distortion of the picture of the animal kingdom. Interspecific and intraspecific relations of non-human animals were then seen mirroring human competitive relations in all their ugly details. The romantic notions of an earlier breed of naturalists and philosophers (e.g., Goethe, Rousseau) were rejected *in toto*. Far from being a peaceful kingdom, animate nature was brutal, selfish, bloodthirsty. For man to be less, to give expression to altruistic impulses, was for him to deny his true nature; this was apparently a common nineteenth-century view (Barzun, 1941).

Businessmen and economists, of course, no less than politicians, rarely took time to observe real animals as distinct from those they imagined prowling the pages of Darwin's *Origin*. Those naturalists who were more concerned with accurate observations than with justification of a particular economic order soon attempted to dispel the "realists'" illusions. Espinas, in 1878, published his *Des Sociétès Animales*. In this work he stressed the near-universal occurrence among animals of social organization and some form of communal life. Far from unrestricted and direct competition being the ruling principle, Espinas found countless examples of social or altruistic behavior. His work should have made an impact, but it apparently failed to do so. As suggested by Allee (1938), the extreme individualism of the late nineteenth-century made notions of co-operative organization repugnant. Just as "nature red in tooth and claw" justified the prevalent mores, a doctrine stressing the universality of social organization appeared to remove that justification. In any event, Espinas failed to convince most of his contemporaries and it is largely in more recent works that we find him cited.

At the turn of the century and in the decade following, a series of studies of social behavior appeared. Some of these were concerned merely with a taxonomy of social forms, as in Deegener's (1918) classification of different levels of social organization. Others sought to prove the existence of a universal social drive or cooperative instinct. The most famous of these last is surely Prince Kropotkin's *Mutual Aid*, published in 1914. As a collection of heartwarming animal anecdotes, it is superb. As a serious contribution to animal sociology, it is somewhat lacking. Kropotkin was as determined to find a biological justification for a benign, co-operative social order among humans as were the economists of an earlier decade to find support for policies of *laissez faire*. Thus, he rather uncritically assembles those tales favorable to his thesis.

Early in the century, Patten (1920) provided formal support for the importance of cooperative principles in biology. Alverdes (1927) later did the same. The latter reviewed the different kinds of social relations to be found among animals and concluded that the overall evolutionary trend was toward a high degree of social interdependence and organization. At the same time there appeared the impressively detailed studies of social life and its evolution among the insects by W. M. Wheeler (1923). Thus, the stage was set for a modern renewal of interest in animal sociology, this time divorced from economic or political considerations. The founder and prophet of this new sociology was, of course, W. C. Allee. Many of his contributions will be noted in the sections following.

The Varieties of Groups

Animal aggregations cannot all be termed societies. When a line of cars moving at high speed along a main road enters a town with reduced speed limits, the vehicles will clump. The clumping cannot be attributed to any social drive of either the cars or their pilots. Similarly, wherever local environmental or microhabitat conditions are such as to cause a reduction in speed of movement, an aggregation will appear. Such aggregations may have survival value, but they require no overt responses to one another on the part of the participating organisms. Nothing is needed, in short, other than a *kinesis* dependent on whatever factors act as the entrapping agent. (For a discussion and example of kinesis and tropisms, see Chap. 8.) In addition to kinetic aggregations, clumping of animals may result from a directional response to a specific environmental condition. If wood lice move toward regions of maximum humidity and if areas of high humidity are localized, aggregations must form. The resulting clumps may be termed *tropistic aggregations*. As with the kinetic aggregations, they need involve no social interactions, merely individual responses to the same environmental stimuli. The selective value accruing to such assemblages, while real enough and perhaps a factor in the evolution of kinesis or tropism to begin with, does not depend on social behavior.

Aggregations may result directly from the response of one animal to another. A male attracted to a female, or parents attracted to their young, or young attracted to one another provide examples of aggregations dependent on animal-centered rather than environment-centered signals. We shall attempt to consider their nature subsequently, although this should not be taken to mean that they operate entirely independently. For instance, in the Gilboa Mountains of Israel herds of gazelles (*Gazella gazella*) graze on slopes of the broad wadis. To the east, visibility extends for several kilometers. There are relatively few visual obstructions. The grass is short, brush or trees are generally confined to a narrow strip in the wadi bottom, and rocky outcrops are few. A few kilometers farther to the west, the wadis are narrower, their sides more precipitous, impediments to distant vision much more frequent. Preliminary observations of the gazelles suggest that these western herds generally move in linear fashion when alarmed. The animals are widely spaced, and distinctive "leadership" is shown by the first and last animals. In contrast, the more easterly herds more often move along a broad front, with no particular animal obviously setting pace or direction. Although our data on these gazelles are as yet incomplete, they do suggest how profoundly sociality and substrate interact.

First, it will be of interest to consider the variety of ways in which these *social aggregations* may be organized. F. Alverdes (1927, 1935) whose energies were largely devoted to a study of animal societies, recognized three categories of relations.

1. First there are mateships which involve a pair of animals whose young disperse, leaving their parents after attaining maturity. Mateships may show varying degrees of exclusiveness and stability. Among some species of bears, for instance, the relationship is monogamous and seasonal. It is also solitary in the sense that the mated pair is not part of a larger aggregate. Alternatively, a permanent though solitary and monogamous relationship may develop, as Alverdes thought was the case with the rhinoceros. Mateships may also be polygynous and seasonal (as with deer) or permanent (zebras, according to Alverdes, 1935).
2. A more complex organization exists when the social unit is the family, a group which may include several adults and the young of previous seasons. In Alverdes's time detailed studies of familiarly

organized mammals were not available. Indeed, what is probably the most complete of such studies dates back no farther than 1955, a study of the black-tailed prairie dog (King), although the thirties and forties did see the publication of other excellent, less inclusive studies (e.g., Carpenter, 1934, 1940).

3. Finally, we come to herds or packs, which are often most elaborate in the complexity of the social relations they exhibit. Herds may include smaller groupings, either families or mateships, hence the increase in the potential diversity of social relations. The red deer, studied in loving detail by F. Darling (1937), have long provided the classical example of an intricately organized herd society. Throughout the winter and early spring the hinds form a society which, in turn, is subdivided into smaller groups consisting of a mature hind and several "followers," generally the offspring of the previous three years. Both in the late spring, at calving time, and again in the fall, during the rut, major realignments occur. Extremes of weather, severe insect infestations, or other disturbances add to the factors that may temporarily shift the organization of what is in essence a highly stable social structure (Darling, 1937).

Social Organization

The organization of animal societies can be attributed to four principal kinds of behavior patterns. These consist of the behavior associated with territoriality and the maintenance of individual distance, with dominance relations, with leadership, and with parental care and mutual stimulation. With the possible exception of the rigidly demarcated insect societies, it is possible to arrange social organizations according to the degree to which one or the other of these patterns predominates. Genetic and nutritional factors play a much greater role in defining the status of the individual insect than is the case with vertebrates. Hence, it is appropriate to consider insect societies as a distinct phenomenon. [For a detailed treatment of insect sociology, see Wheeler's (1923) excellent treatment, that in Allee et al., 1949, and E. O. Wilson's (1971) tome.]

TERRITORIALITY AND INDIVIDUAL DISTANCE

Casual references to the active defense of a certain area by an animal are found in many early natural histories. Credit for the emergence of the concept of *territoriality* in its modern form, however, is generally accorded the English ornithologist, Eliot Howard (1920). He noted that early in the spring the males of many species of songbirds restricted their singing activities to particular trees, shrubs, or fence posts. The areas enclosed by these outposts were rarely if ever traversed by other males of the same species. Thus, Howard concluded, conspecific males divide their breeding grounds into territories each of which is occupied by only a single male. Later on, females, presumably attracted by the advertising song produced by the males as they make their rounds, may join a male in the patrolling of his territory. It will be recognized that there may be innumerable variations in this pattern from one species to the next. These variations may include interspecific differences in the size of the defended territory, the degree to which intrusions by other birds of the same or different species are tolerated, and so on. Sometimes the defended area moves with the individual. For such situations the term *individual distance* has been used (cf. Hediger, 1955), although the basic phenomenon appears to be the same. In either case, a periphery is created at which contact with other individuals may occur.

In cases of extreme territorialism, where each individual or pair has exclusive possession of a tract of land, social behavior will be relatively limited. Absolute dominance by the possessor over any intruder is the rule.

In other cases, such as the gulls, rather small nesting territories are defended, thus forming dense aggregations on the available breeding grounds. In addition to their individual territories, gulls have communal loafing sites which provide opportunity for other kinds of relations than those involved in nest defense. At the loafing site the individual distance shrinks well below the limit obtaining about the nest, for example, and absolute dominance gives way to more fluid inter-individual relations (Tinbergen, 1953a).

It should be noted that an animal's territory is not identical with its *home range*. This last term refers to all of the area in which an animal passes its time, while territory is reserved for that part of the range which is defended against intruders. Although the two may be co-extensive, generally the home range is considerably larger, and the home ranges of many individuals may overlap (for more on territoriality and its evolution, see Klopfer, 1969).

DOMINANCE RELATIONS

Dominance is inferred whenever one individual is able to chastise another with impunity. The ability to be dominant is generally thought to be most frequently a function of sex, size, and physical condition, though by no means is it invariably so. Among the Alaskan fur seals, possession of a female or a harem may be decided in favor of the strongest, heaviest male (so long as he remains the strongest!), but among other territorial beasts the original possessor of a territory may remain the *alpha* or most superordinate animal, almost independently of his physical attributes (see Bonner, 1955, for summary descriptions). The psychological advantage that accrues to the defender of his own home is not unique to *Homo sapiens!*

Dominance is not a constant condition. Among some higher primates (e.g., baboons), for instance, an estrous female assumes the rank of her male consort. Once he loses interest in her, she reverts to her original status, which, in primates, is generally beneath that of the adult males (cf. Washburn and de Vore, 1961). Nor is dominance independent of the context in which it is displayed. An experimental study of dominance among bushbabies (*Galago crassicandatus*) led to the conclusion that the particular index used to measure dominance determined which individual was "dominant." The concept was too heterogenous to be meaningful (Drews, 1973). In a review of the concept and work relating to it, Rowell (unpublished manuscript) flatly rejects it as anything other than a shorthand description of certain behavior patterns. She proposes, instead, that more utility lies in a concept of subordinance.

The notion of dominance as a major phenomenon of social organization is largely due to Schjelderup-Ebbe and its development to W. C. Allee. The former studied chicken societies, where dominance is distributed in a hierarchical fashion (Schjelderup-Ebbe, 1922). The highest ranking animal can dominate all those beneath him, the second highest all but his superior, and so forth. The positions in the hierarchy may be interchanged, of course, as one animal sickens or another grows more vigorous. In other species, such as pigeons, the linear hierarchy is not absolute, giving way to a more changeable peck-dominance relationship such that *A* usually dominates *B*, but not to the extent of being totally immune from retaliatory pecks.

Despite differences in the degree, stability, or nature of the dominance relationship, the establishment of a convention of precedence goes far to assure the transformation of a mere assemblage into an organized society.

LEADERSHIP

Leadership represents an increasingly important phenomenon in mammalian societies whose structure persists over long periods of time, in contrast to the seasonal nature of most bird flocks. By "leader" is meant the individual who determines the direction or rate of movement of a group or who sets its mood, initiating alarm or feeding behavior. Leadership of a group may be divided so that different animals have different roles. Among the red deer (Darling, 1937) the males may act as leaders of their harems during the rut, but should an alarm be sounded, they will depart by themselves, leaving harem leadership to an older hind. Similarly, among reindeer the animals who act as leaders in giving notice of danger are drawn from a peripheral group which does not necessarily include those animals that set the pace (Allee et al., 1949). In both of these cases leadership is fairly stable. Among primates this is apparently also the case, but since leadership there is closely linked to dominance status, a change in the one leads to a change in the other (cf. Washburn and de Vore, 1961). Finally, leadership may in some forms be unrelated either to dominance status, spatial position in the herd, or astuteness. In both schools of fish and groups of ducklings, it is apparently suddenness or directedness of movement that inspires leadership. Whichever animal heads off most determinedly draws the crowd along (Allee et al., 1949). The parallels with human situations are amusing, if not altogether significant.

PARENTAL CARE AND MUTUAL STIMULATION

Among many diverse and unrelated species there are evidences of *parental care* and *mutual stimulation*. The latter includes mutual aid, the grooming of one another by pairs of monkeys (Carpenter, 1940; Zuckerman, 1932), as well as the reciprocal visual and auditory displays seen in birds and fish (cf. Chap. 2). It may also include observational learning, which, as noted earlier (Chap. 4) may be particularly effective in family or close-knit social groups. It is hardly surprising that the prevalence of either parental or reciprocal care or stimulation is so closely associated with complex and long-lasting social bonds. Primate and ungulate societies have perhaps been best studied from this point of view (Yerkes, 1943; Nissen, 1931; Carpenter, 1940; Zuckerman, 1932; Köhler, 1925; Darling, 1937). However, in birds too, where the degree of mutual care or stimulation may vary tremendously from species to species, observations bearing on the same point have been made repeatedly (cf. Friedman, 1935).

Whether most reciprocal responses evolve after a stable social structure has appeared, or, as seems more reasonable, in conjunction with increasingly stable social forms, cannot now be decided.

The Advantages of Sociality

Many of the earlier treatments of animal sociology focused upon the advantages to the organism of the abandonment of a solitary existence (Kropotkin, 1914; Allee, 1938). If such advantages are real and outweigh the disadvantages inherent in social groupings, then one might expect persistent evolutionary trends in the direction of social behavior. Of course, not all organisms are social, or at least not to the same degree. It would appear, then, that just as some factors may favor the evolution of sociality, others disfavor it. After considering the advantages to social behavior we will therefore have to face the question, "Why are all animals not social?"

Even the simplest groups, mere aggregations that result from tropistic or kinetic responses, may produce benefits. Allee (1938) has summarized many such instances. They represent advantages derived from an increased effectiveness of buffers against environmental pressures. A group of animals provide themselves with greater protection against the wind than can one individual. If toxins are present, the larger the aggregations, the greater the dilution effect and the less the likelihood that any organisms will die (Allee et al., 1949). Further, some organisms produce wastes which condition the medium so that its capabilities for sustaining life are improved. Some may grow faster in conditioned water than in the unconditioned, pure substance (Allee, 1938). The conditioning of the medium may be purely physical as well. A larger group of prairie dogs (within certain limits) is likely to fare better than a smaller group because it has more burrows available when a predator threatens (King, 1955). Burrow construction then represents an important form of conditioning which is clearly better performed by several organisms than by isolates. The fact that such advantages do accrue to even the least organized of aggregates will prove of importance when we consider the evolution of societies and social behavior.

In addition to providing a buffer against climatic extremes and promoting growth, aggregates may derive protection from predation. Protection may be a passive affair, as when predators are frightened by numbers of animals which they would eat if encountered individually. The protective effect of aggregation may be the result of an active response of the organisms as well. Though this generally implies a high degree of social organization, it need not invariably do so. Animals could secure protection *en masse* as consequence of the more aggressive behavior of the peripherally placed males. Sea urchins may accomplish the same end by virtue of the physical barrier many spines pose. One of my colleagues described the behavior of a captive seal into whose tank a school of anchovies was released. On previous occasions, when small numbers of the fish were released, the seal rapidly caught and devoured them. The larger number of anchovies, which formed a dense school whose outlines approximated the shape of an animal as large as the seal, not only inhibited pursuit but it also apparently frightened the seal out of its pool. Though anecdotal, this account does illustrate the power of the masses.

The protective values of aggregation are also enhanced because aggregations afford the possibility of polymorphism. A division of labor with certain individuals specializing in defense presupposes the existence of an aggregate. The social insects, of course, provide us with the most striking examples of this phenomenon (cf. Wheeler, 1923).

Passing from the protective values of aggregations, we may consider a second major category of advantages that are related to reproductive functions. First of all, there is the mutually stimulating effect individuals have upon one another which may both synchronize reproductive behavior and facilitate its occurrence. In weaver birds the social stimulus of nest building by some males (itself elicited by heavy rains) results in nest-building behavior by the males throughout the area. Similarly, after the arrival of the females, a certain level of social stimulation during the nest building displays of the males is necessary for successful pair bonding (see review in Farner and King, 1971). It has been proposed (Darling, 1938) that, especially among colonial birds, a population size below a certain minimum will lead to inadequate stimulation and reproductive failure. Above this minimum there is a decrease both in the latency for reproduction and the period of the year over which the breeding season extends (the *Darling effect*). This last could be advantageous in reducing mortality from predators (Darling, 1938). The need for a minimum number of individuals (above two!) and the progressive shortening of a colony's breeding season as colony size increases, has been vigorously disputed (e.g., Fisher, 1954). However, other studies have suggested that it is still a tenable hypothesis (cf. Klopfer and Gottlieb, 1962).

In some cases social grouping may increase feeding efficiency. Among wood pigeons, inexperienced birds seemed to learn the identity and likely location of food from experienced birds during flock feeding. Social feeding behavior also permits the location of sporadically distributed food, which a single-feeding bird might not discover. They also observed that if the food supply is low, solitary individuals have a lower survival rate than do social feeders.

Social patterns can help control mortality by assuring density-dependent controls of total numbers. Consider, for example, that animal populations are capable of exponential growth. When unchecked this growth leads to populations of such a size that depletion of major resources occurs. Thereupon a catastrophic decline in numbers and, depending on local soil and climatic conditions, a permanent demise of the population may occur. The cyclic changes in the abundance of the lynx in North America throughout the nineteenth-century and possibly of the snowshoe hares and lemmings may be examples of such oscillations (cf. Elton, 1942; for recent summaries, cf. Lack, 1954; Keith, 1963). In these cases, the mammalian prey or grasses upon which the animals depend do return, allowing recovery of the population. In tropical lands, however, where denuded lateritic soils may be quickly leached of their scanty mineral reserves, a temporary denudation may not be followed by a recovery of either the flora, or, as a consequence, of the fauna. Under these circumstances, extinction or extirpation will follow. Thus, it is clearly of value for organisms to evolve mechanisms that will limit population growth before such extreme densities are attained. When social aggregates exist such mechanisms can come into play far more rapidly than when individuals live isolated and dispersed. Some of these mechanisms are clearly dependent on social processes. The highly social prairie dogs engage in much contact behavior, mutual sniffing, kissing, and the like. As the number of young increases, their kittenish demands upon their elders lead to a considerable increase in contacts. Ultimately, their elders emigrate, much as a child-besieged father might seek the sanctuary of his secluded office. The peripheral areas to which migration occurs generally lack the well-consolidated system of clearings and tunnels. The emigrants are thus exposed to much heavier predation, which few of them survive (King, 1955). The protection of the habitat has been achieved, however. It is unlikely that these prairie dogs, because of a social organization permitting a control of population density, will ever show the extreme cycles (and thus be exposed to the danger of local extinction) as is the case with northern lemmings (cf. Elton, 1942).

A major study dealing with social behavior and density controls (Wynne-Edwards, 1962) contends that social animals develop conventions which result in reproduction being limited in anticipation of food shortage. The benefit to the group of anticipating responses and reproductive controls is seen as outweighing the individual's loss in fitness which results from a reduction of its reproductive rate below the maximum immediately possible (Wynne-Edwards, 1962). Despite the seeming attractiveness of this notion, not least because it purports to explain the significance of many "displays," it has failed to convince most population biologists (see Williams, 1971). Their principle argument is that if a trait reduces individual fitness, its carriers will disappear, irrespective of any putative value of the trait to a "group." The exceptions lie in those cases in which the members of a group are consanguinous (i.e., have genes in common), for then individual fitness is also related to the survival of the group mates. This is considered below in connection with the evolution of altruism.

Prairie dogs exemplify one other attribute of social animals. The boundaries of the family group's (coterie's) territories are maintained through successive generations. This implies some degree of parental instruction or the existence of traditions. Among primates, particularly man, learned traditions have largely supplanted other means of behavioral control (Yerkes and Yerkes, 1935). The advance in adaptiveness conferred by social organization that allows for tradition learning is considerable. The increased opportunities for learning and for plastic behavior patterns that societies provide must thus be reckoned as one of their most important selective advantages (Yerkes and Yerkes, 1935).

In order to enjoy the advantages of social groupings, animals normally require some form of stable social organization, such as territoriality or dominance. Although the holder of a particular territory may gain advantage from the territory (e.g., by utilizing its food and shelter), a dominant animal in a hierarchy does not consistently gain an advantage. To be sure, the alpha individual may be the first at a food source, and benefit at the expense of lower ranking individuals in the case of a limited amount of food, but it is more probable that the dominance organization benefits all individuals about equally. This is because once the hierarchy is established, no animal need waste time and energy fighting with the others over food or other environmental requirements.

The Evolution of Sociality

We have previously alluded to the fact that all organisms are not equally social. Why not? The list of functions served by social behavior is impressively long. Are there circumstances under which sociality is maladaptive? Or are asocial habits merely due to there having been insufficient time for the occurrence of the necessary evolutionary changes? Affirmative answers are suggested by two studies. Christian (1970) elucidates a case in which, he claims, sociality is maladaptive and has therefore not evolved. His reasoning goes as follows: the habitat of the meadow vole, *Microtus pennsylvanicus*, is wet grasslands, a habitat which is temporary and rapidly passes through successive stages to reforestation. Frequent moves to new areas are necessary for the meadow vole. The high social intolerance shown in the species leads to periodic dispersal whenever population density rises. The dispersal helps assure that a new wet grasslands habitat will be found. Sociality would delay or dampen these irruptions and reduce the likelihood of new habitats being located as the one presently occupied passes into the next seral stage. The prairie vole, *M. ochrogaster*, on the other hand, whose habitat is more stable and extensive, exhibits much less social intolerance than the meadow vole.

Another maladaptive facet of sociality is suggested by a study of wading birds in the British Isles (Goss-Custard, 1970). Although a compact feeding group of waders may afford better protection from predation, this compactness also may interfere with efficient searching for food. In those species in which there is little interference, compact feeding groups and social feeding behavior do develop, presumably in response to the advantage from predation. In those species in which interference is significant, feeding occurs in loosely scattered groups. The degree of scatter represents a balance between the adaptive protection from predators and the maladaptive interference in feeding.

Full answers to these questions are difficult to obtain. Partly because of the difficulty in separating the causes of sociality from its consequences. It is also so because time-consuming and often tedious comparative studies of social behavior are required. Ideally, the animals focused upon should represent all degrees of sociality and should also possess a definite and close phylogenetic relationship. The most favorable organisms for such studies probably are the insects. Their ubiquity, enormous diversity, and the range of conditions under which they occur have made them especially useful for studies of the evolution of social structure. At the same time, of course, the high degree of behavioral stereotypy shown by insects as compared with mammals does limit the range to which extrapolation is possible. In any case, so little is yet known about social behavior and its origins that the temptation to extrapolate has apparently been avoided by most ethologists. Thus, in considering social evolution, we shall have to limit ourselves principally to a discussion of insects.

Most of the many species of insects are, in fact, solitary. A few may be considered subsocial, i.e., they evidence parental care of the young, but no more. O. E. Plath (1935) estimates that less than 3% of the extant species of insects are social or subsocial. Considering the phylogenetic antiquity of many social insect families (early tertiary), this suggests that insect sociality is adaptive only under rather special conditions. Of course, even nonsocial insects may be gregarious at certain times of their lives. Evening swarms of midges (Chironomidae) and the colorful butterfly trees of California are phenomena familiar to most naturalists. Such swarms are only temporary in nature, however, and neither the mechanisms inducing swarming nor the advantages to swarming (except when reproduction is involved) have been studied in any but a handful of cases (e.g., recent studies of locust swarms, Ellis, 1959).

The appearance of stable societies occurs primarily among the Hymenoptera and Isoptera, and to a less frequent degree, among beetles and two other insect orders. Social wasps are mostly found among the 10,000 or so species of the family Vespidae (Plath, 1935; and Wheeler, 1923), although less than 10% of these species are social to any extent. Some of these species do have a rudimentary division of labor and the beginnings of the complex caste systems found in the social bees, ants, and termites. These latter groups (bees, ants, and termites) show social forms of the most elaborate sort. They have been the lifelong objects of study of no small number of naturalists, but foremost among early twentieth-century workers are surely Wheeler (1923), Emerson (1938), and Schneirla (1933).

Wheeler, for whom the origin of sociality was of particular interest, has pointed out that 24 different families of 5 orders of insects show either social or subsocial behavior. The structural similarities between caste members of some of these extant forms and oligocene fossils embedded in baltic amber are so great as to suggest strongly that social behavior goes back at least to oligocene times (Wheeler, 1923). At the same time, some modern insects can be traced back even farther, to the silurian, for instance, so that ancient as social behavior may be, it still remains a recent adaptation for the class Insecta. Since Wheeler sees sociality as promoting the evolution of more highly refined intelligence or adaptability, it is presumably the temporal factor that largely accounts for the absence of sociality in so many species. That is, given more time, many more species of insects should become social. A cogent example of evolving sociality is seen in some sawfly larvae which find it difficult to open pine needles for feeding. As soon as individual feeding sites have been established by those few individuals able to pierce the tough needles, other larvae congregate at these sites and exploit them, too. In groups, larvae have a lower mortality than when isolated. These larvae are interpreted as being on the threshold of permanent sociality with a division of labor (Ghent, 1960).

However, there is accumulating a literature on sociality in vertebrates as well, particularly among birds and primates. A broad summary of much of this work is provided in Crook (1970). The social structure of a species is seen as being broadly adapted to the environment in which the species lives, with ecological factors playing major roles as both ultimate and proximate factors in the social structure of populations. Thus, food supply, predation, and vegetation density act as extrinsic variables impinging upon intrinsic variables of the social structure, such as population size and dispersion rates, modes of communication, and behavior patterns. These intrinsic variables, in turn, exert feedback effects upon themselves and the environment. This system, including the feedback effects, is termed by Crook the "socio-demographic system" of the population.

Crook claims that is has proved possible, especially for the Cercopithecines, to differentiate societies into "grades" which can be correlated with habitat factors. Here are some specific relationships between ecology and sociality which Crook suggests are formed in the "evolutionary theatre":

1. Home-range size of terrestrial open-country primates is a function of food availability in relation to population density.
2. Spacing behavior in arboreal forest primates is a function of high density relative to high-environmental productivity in more or less stable habitats.
3. Social structuring of open-country primates is a function of preda-

tion; coherent groups with organized deployment occur in areas with greatest predation pressure.

4. Group size in open country and forest fringe varies with environmental stability, with a trend to larger groups in areas of medium seasonable aridity.

5. Integration of social units, especially large bisexual troops, varies inversely with density of vegetation and dispersion of food resources.

6. Overcrowding in relation to environmental supply of commodities results in increased male exclusion from reproductive groups; this tends to produce a diaspora of solitary males in a forest and all-male groups in open country.

The Ontogeny of Sociality

Collias (1944, 1951) proposed dividing the stages of socialization into categories which one can consider separately. This is a useful approach if we do not allow it to obscure the multifactorial and interlocking nature of the process of socialization.

AN INITIAL PREDISPOSITION TO RESPOND TO OTHER ORGANISMS OF THE SAME SPECIES

Two puppies placed together will be playing with one another in short order, a pair of hens will likely peck at one another, but two frogs (outside of the breeding season) will probably show no signs of mutual recognition. In this last instance, there is little in the way of social responses to build upon for the development of any sort of social structure, just the reverse of the example afforded by the puppies. This is not to suggest that the initial predisposition to make social responses is in every way independent of experience. As we shall see later, it is not. But, as can be seen in the examples cited by Plath (1935), even among closely related groups such as the insect orders, tremendous differences exist with respect to initial responsiveness to conspecifics.

SELF-REINFORCEMENT OF INITIAL RESPONSES

The second phase of socialization, according to Collias (1951), *autoreinforcement*, has been the subject of much more study, albeit most of it of a recent nature. As an example of this phenomenon, consider the work of Mavis Gunther (1961); in human mothers, breasts of a certain shape and size cause the upper lip of a nursing infant to be rolled back against its nostrils, interfering with breathing. This provokes a violent withdrawal on the part of the infant. One or two nursing sessions are often all that is required before merely bringing the infant toward the breast provokes crying and rejection responses. Imprinting (see Chap. 2) represents a more subtle form of reinforcement in which a response (following) itself strengthens the probability of its maintenance.

INCREASING SOCIAL DISCRIMINATION

Following Collias, we can recognize a third phase in ontogeny when the young animal begins to discriminate between individuals toward whom different kinds of responses must be directed. Males and females, juveniles and adults, dominants, peers, and subordinate animals are not usually treated identically. Many of the distinctions are learned by trial and error. Even a playful puppy can quickly learn to retreat upon evoking an especially grim growl from an adult. Beach's studies of sex discrimination emphasize the role of experimental factors in social discriminations (Beach, 1947; Kagan and Beach, 1953). He found that young male rats reared with mature females failed to develop normal sexual responses. A pattern of play activity associated with females had apparently conditioned the young males so that their recognition of females as sex-objects was barred.

Much of Lorenz's work has been concerned with this phase of socialization. He suggests that, at least among birds, social discriminations depend on imprinting or imprinting-like processes. These can only occur at specific periods during the animal's life (*critical periods*), so that the ability to make the normal social discriminations depends largely on the presence of the appropriate individual(s) at a specific time (see Chaps. 2 and 12; Schutz, 1965).

SOCIALIZATION GUIDED BY FAMILY

For many animals, the processes of socialization end with learning to discriminate among individuals of different rank and sex. Those with extended familial relations, however, may develop more complex modes of social behavior under the guidance of the parents or family group. Among the higher primates and man this is most pronounced, but other forms partake of the benefits of family traditions as well. One of the best chronicled of these is the black-tailed prairie dog cited earlier (King, 1955). A lengthy association between juveniles and adults serves to acquaint the former with the limits of the coterie territory and the social signals (grooming, kissing, etc.) to maintain the coterie bond.

SOCIAL INDEPENDENCE AND REINTEGRATION

In time, many family groups break up, with the adults and the young going their separate ways. This introduces the period of social independence. It is not necessarily a period of solitude, however. Among deer, the young males are chased from the hind and fawn groups in the fall and band together (cf. M. Altmann, 1952; Darling, 1937). Only a rather brief period of solitude exists, followed by the formation of juvenile clubs. In species in which the family is the only social group, the period of solitude may persist until the young find mates and commence a family of their own. In either case, this phase soon gives way to the next, that of reintegration. The degree to which reintegration is influenced by previous experience or other factors has only recently begun to be a topic of major interest to ethologists. Clinical and social psychology have, of course, long had to deal with this problem. The ethological studies bearing on this point have largely entailed separations of infants and mothers at various ages or the provision of surrogate parents. This work is considered below.

Some Recent Developments

As was suggested earlier, most of the more recent advances in our understanding of animal sociality concern the ontogeny of social behavior, though interest in the evolution of social behavior remains alive too, as evidenced by a paper by Hamilton (1963) on altruism. Other new developments have occurred in studies of communication and the definition of social roles.

In general, altruistic behavior (behavior that benefits another while exacting a price from the performer) is dependent on the existence of some form of social organization. However, acts which benefit the recipient at the expense of the donor would not likely be perpetuated unless certain conditions are met. Specifically, for selection to favor altruism the advantage to relatives of the altruist must exceed the individual disadvantage. For siblings of the altruist, the gain, for instance, must equal at least twice the loss, e.g., two siblings' lives must be saved for the death of one altruist. For more distant relatives, a correspondingly greater benefit is required. When such conditions are met, altruistic behavior can be expected to evolve.

The behavior of the yellow-faced grassquit (*Tiaris olivacea*) apparently ranges from social and nonaggressive to territorial, but on the island of Jamaica it is very aggressive (Pulliam et al., 1972). The probable reason was examined in a study that illustrates a growing interest in an empirical approach to what have been speculative issues.

An aggressive territory holder can decrease the fitness of a non-aggressive bird by excluding it from optimal habitat. It is less obvious that the decrease in fitness of the nonaggressive bird is greater than the increase in fitness of the aggressor. However, the territorial bird does lose some of the advantages of social behavior (whatever they are) and must spend considerable time defending his territory, time which might otherwise be applied toward maintenance and reproduction. The amount of time which the average aggressive individual spends defending his territory must necessarily increase as the proportion of the bird population which is territorial increases. Hence, the question: why are some grassquits territorial?

Suppose territorial individuals do have a lower reproductive capacity than social individuals would have in the absence of the former. This would result in a territorial population maintaining lower numbers than a social population even though the territorial individuals were superior in competition with the social individuals! If, for a given bird species, the social populations were shown to maintain a significantly higher population density than the territorial populations, we would have evidence that territoriality is a selfish behavior for that species (Pulliam et al., p. 78).

Pulliam (1973) censused, during the breeding season, 11 similar habitats that appeared suitable for yellow-faced grassquits in both Jamaica and Costa Rica. Each habitat was visited twice. In Costa Rica, on a total of 25.9 acres, an average of 20.5 grassquits were seen. In Jamaica, on a total of 18.0 acres, an average of only 6.9 grassquits were seen. In both Costa Rica and Jamaica, there were grassquits in 4 of the 11 habitats visited. The number of grassquits per acre in those sites containing some grassquits was 2.9 in Costa Rica, as compared to 0.7 in Jamaica. The increase in the density of the Costa Rican grassquits is especially surprising since there were many more individuals and species sharing sites with grassquits in Costa Rica than there were in Jamaica. Thus, it appears that the social grassquits of Costa Rica are able to maintain a population density two to three times as great as that of the territorial Jamaican grassquits. This accords with our supposition.

Very little is known about the degree of heterozygosity in natural populations of birds and we are not yet able to predict the degree of heterozygosity that might permit selfish traits to evolve. However, we do know that both isolation and population size exert considerable influence on the degree of genetic diversity of natural populations. In very small populations, random actual censuses of the populations, of the resources available to them, and of their degree of consanguinity are allowing a test of the model.

With regard to the ontogeny of social behavior, Harlow and Harlow (1962) have described the role of parental attention on the development of young rhesus monkeys. They were able to show that much of the re-inforcement and other effects of maternal rearing are independent of satisfaction of *primary drives:* hunger, thirst, warmth, and contact. Inanimate wire and cloth dummies, though adequate sources of food, support, and warmth, still cannot provide baby monkeys with certain other ingredients, what the Harlows term *mother love.* Without these the monkeys eventually become neurotic and asocial caricatures of normal young or else show severe disturbances in their later maternal behavior (Harlow and Harlow, 1962). Similarly, opportunities to play must not be denied else specific distortions in behavior may result (Müller-Schwarze, 1969; Chepko, 1971). These studies have been followed by a great many others dealing with a large array of primate (and some other) species. The results do not uniformly support the conclusions reached by the Harlows and other students of rhesus monkeys. This, of course, is entirely consistent with the concept of a multifactorial control of behavior, as is implied, for instance, in the review of primate sociality by Eisenberg et al. (1972). Generalizations on the ontogeny of sociality (and other behavior) can thus not be based on one, two, or three "representative" species (cf. Klopfer, 1973). This point is considered at greater length in Chap. 10.

The meanings of visual or auditory signals may also fail to become clearly established if there is early social deprivation. Hand-reared fallow deer, for example, apparently do not later become integrated into a normal herd as a consequence of their being reared apart from adults of their own kind because the "meanings" attached to certain movements are misunderstood (Gilbert and Klopfer, personal observation).

The field of animal communication in the larger sense has begun to expand rapidly. Earlier studies were limited to either catalogues of signals (ethograms) and the responses attached to them or to the sensory basis for signal reception and transduction. The publication in 1968 of *Animal Communication* (Sebeok, ed.) symbolized a change. The collection of papers contained within the volume can hardly claim to represent any theoretical or empirical breakthroughs; indeed, most are pedestrian, if useful, compilations of facts and old theories. The symbolic significance lies in the formal announcement the book provides of a marriage between linguists and biologists. Their alliance, in fact, has grown into a true group marriage, with mathematicians (Cohen, 1969), neurophysiologists (Pribram, 1971), anthropologists, and systems analysts, *inter alia* (Sebeok, 1968). The term *semiotics* has been adopted as a label for this field, which is defined as "the study of patterned communication in all modalities" (Sebeok, 1968). The central issues to which much of the current work seems to be directed concerns the holistic or gestalt nature of signs and signals, i.e., the dependency of meaning on context, and the ontogeny of communicative abilities. The pre-Darwinian interest in the phylogeny of communication and in the neural mechanisms that underlie language are two other areas that continue to receive attention.

Finally, the issue of role assignment has become the subject of experimental studies. The issue has been raised in two ways:

1. Are "roles" independent of context; i.e., is there a "dominant" animal that is able to be superordinate in every situation?
2. Is dominance, i.e., priority of access, specific to particular situations?

As mentioned above, a study of bushbabies (Drews, 1973) asserts the latter. A review by Rowell altogether denies the usefulness of the concept of dominance, although this does not necessarily deny that there may exist other roles which are independent of situational context.

The other issue concerns how the "leaders" come to their posts. Are roles assigned according to genetic heritage? Are they derived from or influenced by the behavior of parents? Rhesus monkey males apparently attain a rank related to that held by their mothers (Koford, 1963; Sade, 1968). Two recent studies have focused upon intra-litter differences in the behavior of young wolves (Fox, 1971) and goat kids (Klopfer and Klopfer, 1973) and their relation to subsequent adult roles. These studies suggest a far more complex answer than that implied by the original questions on the importance of heritage, parentage, and early experience. This is explored further in Chap. 12.

Public interest in the field of mental health has greatly intensified interest in animal social behavior (cf. Josiah Macy *Proceedings*, 1954–1955, B. Schaffner, ed.). Thus, we may expect an ever-increasing amount of work devoted to an elucidation of the processes of socialization. Particular attention will likely be given specific experimental factors that facilitate social behavior of various kinds (Klopfer, 1961, 1973): endocrine factors affecting aggressiveness, inherited predispositions (strain differences) for social behavior, and maternal influences (including *in utero* effects) (Hirsch, 1962; Denenberg and Whimbey 1963; Newton and Levine, 1968; Foss, 1965). Much of this work dating to the period 1950–1970 has been reviewed by Altmann (1960), Ambrose (1969), Esser (1971), Scott (1956), Wynne-Edwards (1962), and in a volume edited by Rheingold (1963).

Overview

Nineteenth-century studies of social behavior generally reflected a philosophic bias. Animate nature was seen by some as brutish and competitive, by others as kindly and cooperative. With the end of that century, the emphasis on co-operation as the governing principle of social evolution grew still stronger: Kropotkin's (1914) *Mutual Aid* has been the solace of many a pacifist (Klopfer, personal observation). In Kropotkin's volume one can find a biological rationalization for an ordered and peaceful society.

Another decade passed before more objective and systematic studies of social behavior became popular, foreshadowed by Patten (1920), then Alverdes (1927), Wheeler (1923), and culminating with Allee (1938). Under the leadership of these men, detailed descriptions of the varieties of groups began emerging, from the insects (Wheeler, 1923) to the primates (Carpenter, 1934; Nissen, 1931; Zuckerman, 1932). The concept of territoriality introduced by Eliot Howard (1920) led to discussions of specific mechanisms of social organization, including the important notions of individual distances and dominance. Finally, Allee (1938) and his students were able to provide experimental support for the belief that social behavior confers selective benefits and promotes survival.

The last two areas to which we turned our attention, the evolution and ontogeny of sociality, still represent largely unplowed ground. Wheeler (1923) and his successors have been able to attempt phylogenies of behavior among certain insect groups, but the larger question concerning the reasons for the occurrence or nonoccurrence of social behavior has yet to be convincingly framed. Similarly, in the case of the development of social behavior, it is now possible to implicate genetic, intra-uterine and experiential factors, but a satisfactory theoretic framework still awaits erection.

Selected Readings

The best general discussion of the earlier studies of social behavior is probably that by W. C. Allee (1938). A more systematic review of the subject is *Principles of Animal Ecology* by W. C. Allee, A. E. Emerson, A. Park, T. Park, and K. Schmidt (Saunders, 1949). Specific material on insect, avian, and mammalian sociality is available in C. Murchison's *A Handbook of Social Psychology* (Clark Univ. Press, 1935) and (for insects) in E. O. Wilson's *The Insect Societies* (Belknap, 1971). Tinbergen's *Social Behavior in Animals* (Methuen, 1953) approaches the subject from the viewpoint of classical ethology. The anthology by Rheingold (1963) brings together much work on the influence and development of maternal behavior; and more current summaries are to be found in Ambrose (1969), Foss (1965), Newton and Levine (1968). Three primary sources for material on sociality and space are Esser (1971), Sommer (1969), and Jewell and Loizos (1966). That last volume concerns itself with territorial and explorative behavior, as well as play. Material on semiotics is best sought in Sebeok's (1968) collection and R. A. Hinde's *Nonverbal Communication* (Cambridge Univ. Press, 1972).

Chapter 8

THE
ORIENTATION
OF
ANIMALS

Tropisms, Time-Compensated Orientation and Navigation

Introduction

In this chapter we shall turn our attention to what were originally three separate lines of inquiry: studies of directed movements (the tropisms, taxes, and kineses of various authors); of biological rhythms and chronometry; and finally, of homing and migration. At the present time, our realization of the close interrelation between these facets of animal behavior virtually precludes considering them in isolation from one another. Nonetheless, the earlier studies were sufficiently distinct to allow separate treatment here.

Directional Movements

Even before Darwin's studies on the subject (*The Movements and Habits of Climbing Plants*, 1876), publications on directed movements of both protozoans and plants had become available. In addition, descriptions of geotropisms, or movements in response to gravitational forces (Knight, 1806), and phototropism, or similar responses to light (de Candolle, 1832), had appeared in the scientific press. During the years throughout the latter half of the nineteenth century, various authors described other directed movements on the parts of plants, protists, and bacteria in response to chemicals, moisture, pressure, temperature, and other stimuli. (A summary of this early work will be found in Fraenkel and Gunn, 1940, and Mast, 1938.) The most important synthesis of the ideas advanced in these studies was produced by Jacques Loeb in 1890 and 1906 (also see Chap. 1). In brief, Loeb (1906) believed that whenever a source of stimulation impinged asymmetrically upon the sense organs of a bilaterally symmetrical creature, a similar inequality in the magnitude of muscle contraction on the opposite sides of the animal would result. The effect would be to continually alter the position of the animal until its sense organs became equally (i.e., symmetrically) stimulated and the bilateral muscle groups equally active. This was considered an accidental process and of little, if any, adaptive value (Loeb, 1918: 13–16):

> Normally the processes inducing locomotion are equal in both halves of the central nervous system, and the tension of the symmetrical muscles being equal, the animal moves in as straight a line as the imperfections of its locomotor apparatus permit. If, however, the

velocity of chemical reactions in one side of the body, e.g., in one eye of an insect, is increased, the physiological symmetry of both sides of the brain and as a consequence the equality of tension of the symmetrical muscles no longer exist. The muscles connected with the more strongly illuminated eye are thrown into a stronger tension, and if new impulses for locomotion originate in the central nervous system, they will no longer produce an equal response in the symmetrical muscles, but a stronger one in the muscles turning the head and body of the animal to the source of light. The animal will thus be compelled to change the direction of its motion and to turn to the source of light. As soon as the plane of symmetry goes through the source of light, both eyes receive again equal illumination, the tension (or tonus) of symmetrical muscles becomes equal again, and the impulses for locomotion will now produce equal activity in the symmetrical muscles. As a consequence, the animal will move in a straight line to the source of light until some other asymmetrical disturbance once more changes the direction of motion.

What has been stated for light holds true also if light is replaced by any other form of energy. Motions caused by light or other agencies appear to the layman as expressions of will and purpose on the part of the animal, whereas in reality the animal is forced to go where carried by its legs. For the conduct of animals consists of forced movements.

The idea that the morphological and physiological symmetry conditions in an animal are the key to the understanding of animal conduct demanded that the same principle should explain the conduct of plants, since plants also possess a symmetrical structure. The writer was able to show that sessile animals behave toward light exactly as do sessile plants; and motile animals like motile plants. The forced orientation of plants by outside sources of energy had been called tropisms; and the theory of animal conduct based on the symmetrical structure of their body was, therefore, designated as the *tropism theory of animal conduct.*

This view was in accord with that of several earlier workers, notably Verworn (1889).

Some of the dissent from Loeb's view may be considered due to philosophic objections to mechanistic theories *per se* (e.g., Fabre, Chap. 1). Other dissenters merely objected because they had found particular examples of directed movements that could not be fitted into Loeb's scheme. The most vigorous and reasoned dissent, however, came from S. O. Mast (1938). Along with other workers. Jennings (1906), Holmes (1905), von Buddenbrock (1915), and Kühn (1919), Mast (1938) showed that the processes of orientation differ among various groups of animals and do not fit into Loeb's simple muscle-tonus theory. Among rhizopods, orientation results from a "gelating effect of light" (Mast, 1938) which inhibits pseudopod formation on the illuminated side. Jennings (1906), in turn, had shown that much protozoan movement was not even related to the direction or source of the stimulus. In annelids, orientation is largely the result of random movements, with symmetry of form being largely inconsequential. Insects depend on a complex series of reflexes for orientation. The main point Mast (1938) stressed was that proper orientation represents an adaptive condition. Hence, natural selection would favor *any* mechanisms assuring appropriate orientation, and not merely the single muscle-tonus system Loeb believed to be a universal mechanism for orientation. To quote Mast (1938: 218–219):

> There are two fundamentally different types of photic response in animals and both are found in nearly every animal that responds to light. The one is quantitative. It consists primarily of increase or decrease in the rate of activity, correlated with the amount of light absorbed by photosensitive substance. It may be designated a kinetic response. The other is largely qualitative. It consists primarily of change in direction of movement, correlated with the rate of change in the amount of light absorbed by the photosensitive tissue. It may be designated a shock-reaction.
>
> Both types of responses usually result in bringing the animals involved, especially the lower forms, into and keeping them in regions which are favorable in reference to environment. For example, some animals get into favorable regions by random movements (trial and error) and remain there because they become less active, i.e., owing to kinetic responses (turbellaria); others get into these regions by random movements and remain there, owing to shock-reactions induced by environmental changes at the boundary (*Euglena*); some get into the regions by orientation usually brought toward these regions and remain there, owing to reversal in the direction of orientation in supra- and suboptimum light intensities (*Volvox*); others get into the regions by movement directly toward them, even if orientation is prevented, and remain there, owing to the action of a number of interrelated factors associated with "home" (ants).
>
> All three methods function in regulation of the environment by locomotion, but the first is the simplest and doubtless the most primitive. It depends merely upon correlation between activity in the locomotor mechanism and the intensity of the light. It is a purely "trial and error" method. The second is more complicated but obviously much more efficient in attaining favorable environmental conditions. It usually depends upon the action of a directive stimulating agent upon differentiated receptors. The third involves the ability of using various means to attain an end so that if one fails others can be substituted. It is consequently more efficient than either of the other two.

It is common among higher animals, especially in man, but it is probably also found in lower animals, as indicated by the change in the factors involved in the process of orientation in *Planaria* after one eye has been removed and in *Eristalis* after some of the legs on one side have been removed. This is particularly obvious in the latter in which, after the operation, the action of the remaining legs changes radically so as to compensate for the loss of the action of those which have been removed. As a result of this compensation the insect proceeds toward the light somewhat sidewise in place of directly forward. That is, if *Eristalis* is prevented from going toward the light in the way which usually obtains, it adjusts its responses and goes toward it in another way. This indicates that the attainment of an end in view is involved.

Orientation is consequently one of several phenomena involved in the regulation of the environment of animals by motor responses. It is intermediate in complexity and in efficiency and it probably evolved from simple types of response in accord with the views of Jennings (1906).

In short, while Loeb attributed most, if not all, orientation responses to quantitative differences in stimulation of symmetrical receptors and effectors (the muscle-tonus system), Mast and his school were not willing to accept a single mechanism for all organisms. If orientation abilities were adaptive, as Mast believed them to be, many different orienting mechanisms could be expected to have evolved. Some support for Loeb's point of view in this controversy has come from Crozier and his students (Crozier and Hoagland, 1934), though in general Loeb's position has found increasingly fewer advocates. In 1919, a classification of orienting responses was proposed by Kühn, which, had it only received a wider audience, might well have mitigated the Loeb-Mast debate. Kühn's classification formed the basis of what still remains the most complete summary of directed movements, that of Fraenkel and Gunn (1940). The similarity between the latter's scheme and that of Kühn is so great that we may turn directly to the outline and definitions proposed by Fraenkel and Gunn and reproduced in Table 8–1.

Fraenkel and Gunn's (1940) aim was primarily to identify and classify the various kinds of orienting responses characteristic of animals. In doing so, they could not but refute Loeb, for they uncovered a multitude of examples in which organ symmetry was irrelevant for orientation. At the same time, they recognized the dual meaning implicit in Loeb's theory and made an important distinction between its two constituent elements.

Loeb's theory, which is variously referred to under the rubrics of "muscle-tonus," "tropisms," or "forced movements," implies first of all, that there is no freedom of choice or will in animal movements. In the context

TABLE 8–1

1 GENERAL DESCRIPTION	2 FORM OF STIMULUS REQUIRED	3 MINIMUM FORM OF RECEPTORS REQUIRED

KINESES. Undirected reactions. No orientation of axis of body in relation to the stimulus.

Ortho-kinesis Speed or frequency of locomotion dependent on intensity of stimulation.	Gradient of intensity	A single intensity receptor
Klino-kinesis Frequency or amount of turning per unit time dependent on intensity of stimulation. Adaptation, &c., required for aggregation.	Gradient of intensity	A single intensity receptor

TAXES. Directed reactions. With a single source of stimulation, long axis of body oriented

Klino-taxis Attainment of orientation indirect, by interruption of regularly alternating lateral deviations of part or whole of body, by comparison of intensities of stimulation which are successive in time.	Beam or steep gradient	A single intensity receptor
Tropo-taxis Attainment of orientation direct, by turning to less or to more stimulated side, by simultaneous comparison of intensities of stimulation on the two sides. No deviations required.	Beam or steep gradient	Paired intensity receptors
Telo-taxis Attainment of orientation is direct, without deviations. Orientation to a source of stimulus, as if it were a goal. Known only as response to light.	Beam from a small source of light	A number of elements pointing in different directions

TRANSVERSE ORIENTATIONS. Orientation at a temporarily fixed angle to the direction is seldom directly toward or away from the source of stimulation.

Light compass reaction Locomotion at a temporarily fixed angle to light rays, which usually come from the side.	Beam from a small source	A number of elements pointing in different directions
Dorsal or (ventral) light reaction Orientation so that light is kept perpendicular to both long and transverse axes of the body; usually dorsal, but in some animals ventral. Locomotion need not occur.	Directed light	Paired intensity receptors
Ventral earth (transverse gravity) reaction Orientation so that gravitational force acts perpendicularly to long and transverse axes of body. Dorsal surface usually kept uppermost. Locomotion need not occur.	Gravity (or centrifugal force simulating gravity)	Statocysts with a number of elements, as is usual with statocysts

After D. S. Fraenkel and D. L. Gunn, 1940, *The Orientation of Animals*, Oxford Univ. Press.

4 BEHAVIOR WITH TWO SOURCES OF STIMULATION	5 RESULT OF UNI-LATERAL REMOVAL OF RECEPTORS	6 FORMERLY CALLED	7 EXAMPLES*
Locomotion random in direction.			
Reaction to whole of the gradient	No effect? Reduced intensity of reaction?	Simply kinesis	*Porcellio*; Chap. II
Reaction to whole of the gradient	No effect? Reduced intensity of reaction?	Phobo-taxis; avoiding reactions; *Unterschiedsempfindlichkeit*	*Dendrocoelum, Paramecium*? Chap. V
in line with the source and locomotion toward (positive) or away from (negative) it.			
Orientation between the two, curving into one when close to sources See Chap. XI*	Usually impossible; no effect? Reduced intensity of reaction?	Part of tropo-taxis; avoiding reactions; phobic mechanism	Fly larvae, *Euglena*, larvae of *Arenicola, Amaroucium*; Chap. VI
Orientation between the two, curving into one when close to sources See Chap. XI	Circus movements in uniform field of stimulus and often also in beam or gradient	Tropo-taxis, but klino-taxis excluded, and cases with two-way eyes added from telo-taxis	Woodlice, *Ephestia* larvae of *Eristalis, Notonecta*; Chap. VII
Orientation to one at a time; animal may switch over to the other at intervals, giving a zigzag course	Not known, because often the same animal can behave tropo-tactically, and the circus movements which occur may be the result of tropo-taxis	Telo-taxis, but excluding cases which always orientate between two lights before curving into one of them, and excluding reactions to specific objects and form reactions	*Apis, Eupagurus*; Chap. VIII
of the external stimulus or at a fixed angle of 90°. Locomotion need not occur and in any case			
If each light affects one eye only, orientation is to one alone; if one eye is affected by both lights, orientation to both combined	No effect, except in limiting the possible angles of orientation	Meno-taxis	*Elysia*, ant, bees; caterpillars of *Vanessa urticae*; Chap. IX
Not known	No effect in some species; produces screw-path in others	Dorsal light reflex (*Lichtrückenreflex*)	*Argulus, Artemia, Apus*; Chap. X
A statocyst combines the forces mechanically and orientation should be determined by the resultant	No effect in some species; produces rotation or lateral tilting in others	?	*Leander*, Crayfish; Chap. XV

*Chapter references are to Fraenkel and Gunn, 1940.

of modern biology, this is hardly a meaningful assertion. If free will does not exist, animals of identical heredity and under identical conditions should respond in an identical fashion. But, as Fraenkel and Gunn (1940) point out, uniformity in responses could also be ascribed to the animals "wanting" to do the same thing. Alternatively, if there was variation in the responses of different animals, this could be due to free will or to the fact that conditions were not absolutely identical. In short, the question of free will cannot be resolved by experiment. However, to quote Fraenkel and Gunn (1940: 308): "As long as there is order in nature, we can fruitfully investigate it experimentally. Statistically speaking, even free will is orderly and is not beyond investigation; we can get on quite well without knowing whether it exists or not." However, at the time Loeb first proposed his tropism theory, anthropomorphic attitudes were still sufficiently common (cf. Chap. 1) that his efforts to interpret behavior in physical terms deserve applause.

The second element of Loeb's theory concerns the mechanism of the orientation process. It is here that stress is placed on the symmetry of receptor and effector activity. And it is here, too, where the shortcomings and rather restricted applicability of what was intended as a universal explanation become apparent. Reference to the table, for instance, will show that light compass reactions represent oriented locomotion during asymmetrical stimulation. Fraenkel and Gunn (1940) also cite experiments in which orientation was not impaired either by the removal of a limb from one side or by other artificial interference with normal body symmetry.

The modern successor to the mechanistic approaches described above is seen in the application of control theory (*cybernetics*) to directional orientation. Consider the orientation of a spider to a potential prey organism. When one of its four lateral eyes detects an object, it turns toward it, just the right amount, so that the object will fall into the field of the frontal eyes. The amount of the turn is specified in advance and is independent of visual feedback. Is the command to turn based on a specific turning time? Apparently not, since the speed of turning may vary. Nor is the proprioceptive input from the legs necessary. However, since the angle through which the animal turns with each step is constant, the turning command may involve the computation and specification of a discrete number of steps (Land, 1972). A review of insect control systems and their general features is provided by Mittelstaedt (1962a).

Biological Rhythms and Chronometry

"Rhythmicity is characteristic of nature" (Cloudsley-Thompson, 1961), whether we consider sidereal motions, the alternation of the seasons, or the pumping of the heart. Scientific attention was first focused on animal rhythms at the beginning of this century. Von Buttel-Reepen (1900) and Forel (1910), observing the regularity with which bees sought food at a particular locality (e.g., Forel's breakfast table), concluded that bees possessed a time sense. This, in turn, implied (though it did not prove) the existence of a stable interval cycle or rhythm, and this was probably the first rhythm to be seriously studied. Such rhythms, in turn, are probably fundamental to cyclically recurring behavior such as diurnal activity cycles, as well as to the complex pattern of responses involved in animal migration and navigation.

The study of biological rhythms, their functions and controlling mechanisms, is an absorbing branch of biology. In order to make a clear assessment it will be well to begin by contrasting the various ways in which cyclic biological phenomena have been catalogued. We can then distinguish four different classificatory schemata. The first of these describes biological rhythms or cycles in terms of their frequency (i.e., whether annual, daily, or hourly). The second is concerned with the overt expression of the rhythm (i.e., whether it is a change in a specific metabolic function or in gross behavior). The third depends on the adaptive function of the rhythm. The fourth, finally, is based on the controlling stimuli or centers and the nature of the basic timing mechanism. This last, of course, carries one deep into the realm of cellular physiology and biochemistry.

CYCLES IN TERMS OF FREQUENCY

Kleitman (1949) has written a "review of reviews" in which are cited most of the important earlier papers on periodicity. He distinguishes, first of all, periods of less than 24 hours. These include two-hour cycles of activity or feeding in laboratory rats and other rodents and four- to five-hour cycles in human infants. These cycles are linked with periodic changes in gastric motility and hunger. In a salamander there were found marked changes in thresholds for tactile stimulation occurring at eight- to ten-hour intervals. Rhythms such as these Kleitman terms *polycyclic*, a term meant to include any function that has a period of 12 or less hours. This replaces an older term, polyphasic, introduced by Szymanski (1914), another of the pioneers in the studies of rhythms.

"Diurnal periods," in contrast to those that are polycyclic, occur but once in 24 hours. The term "diurnal," however is misleading in this context since many animals with 24-hour activity cycles are maximally active at night. Properly speaking, theirs are "nocturnal" periods. Further, few 24-hour cycles are held precisely at 24 hours. As we shall see subsequently, they usually range from 23 to 25 hours, though any one individual may show an extraordinary constancy in his own period. Hence, diurnal period has come to be replaced by *circadian rhythm* (*circa*, about; *dies*, a day). Circadian rhythms appear to exist for almost every animal or function in which they have been sought. Examples, as well as some of the reasons for this, will be considered subsequently.

Periods of more than 24 hours include estrous and sexual activity cycles, which vary from 4½ days (laboratory rats) to 12 months (in seasonally breeding birds, mammals, fish, and insects). Also included here must be the six-day cycles in the feeding of some insects and the phosphate content of rabbit muscle (Kleitman, op. cit.).

At this juncture it is appropriate to point out that many allegedly cyclic phenomena represent statistical artifacts. Nowhere is this point more cogently and humorously made than in an article by Lamont Cole (1957). His "study" (of a table of random numbers) demonstrated the existence of a rhythmic change in the metabolic rate of a unicorn! The fact that the mathematical techniques used in this demonstration are often used by workers in the field of chronometry and cycles should induce a healthy measure of skepticism toward many reported "cycles." Consider such claims as that eminent people tend to be born in greatest number in winter or that revolutions are cyclic, following the sun-spot cycles (cited by Kleitman, 1949).

CYCLES IN TERMS OF OVERT EFFECTS

General reviews dealing with the different sorts of physiological cycles that have been the focus of attention will be found in Welsh (1938), Kleitman (1949), and Brown (1957) (also cf. Aschoff, 1962; Bünning, 1963; *Cold Spring Harbor Symposium on Quantitative Biology*, 1960; Menaker, 1971). The examples that follow are taken from these reviews.

Since 1900, pigment movements, particularly in the eyes and exoskeletons of arthropods have been known to be cyclic. Presumably, such movements are related to diurnal changes in illumination, although they may proceed independently of such changes. Within the photo-receptors of the eye, cyclic movements of pigments are now known in noctuid moths, several genera of crustaceans, and fish, among others. Cyclic color changes of the entire body are also known to occur in some amphibia and fish.

Periodic changes in activity or physical restlessness and in metabolism have also been regularly observed since early in this century. Such activity cycles of varying frequency and stability are known for virtually every order of plant and animal yet examined, from mammals on down. Body temperature can be shown (in mammals and birds) to cycle similarly.

Emergence or hatching rhythms are known for some insects, while birth rhythms in mammals are also occasionally claimed. These latter claims are not always convincing.

For the earlier stages of reproduction, the existence of some rhythms is indisputable. Spermatogenesis, ovulation, behavior associated with reproduction (courtship, nest building, etc.), all can be shown to vary both diurnally and annually.

CYCLES IN TERMS OF ADAPTIVE FUNCTION

There are certainly cycles whose adaptive features have yet to be discovered. It is also possible that some cycles may represent incidental by-products of other processes and are in themselves of neither value nor harm, analogous to the pleiotropic effects of certain genes. However, some suggestions as to the kinds of benefits which can be provided by cycles will be gleaned from the following list.

Annual cycles of reproduction, where these are timed by environmental agents of widespread or universal occurrence, assure simultaneity in male and female sexual behavior. In most temperate-zone birds and some mammals, the external timer for this reproductive rhythm is a particular day length (cf. review in Wolfson, 1959). Thus, all members of a particular species residing at a particular latitude will show phase synchrony with respect to the reproductive processes. Diurnal or polycyclic rhythms of reproduction have the same effect.

Daily activity cycles are of benefit in as much as they may aid an animal in avoiding predators. (Of course, a nocturnal animal's predator may also evolve efficient means for nocturnal hunting; in this case the benefit is the predator's!) In areas where birds are hunted daily except Sunday, a seven-day cycle of tameness has been noted (Meinertzhagen, 1950). Nocturnal or cyclic activity may also provide less competitive feeding opportunities, especially if the cycles of the most similar sympatric species are out of phase with one another. Protection from extremes of temperature or water loss is essential to many desert-dwelling forms. This protection is often afforded by a subterranean habit, emergence being timed to coincide with the setting of the sun (Cloudsley-Thompson, 1961, Chap. 3).

In animals of the seashore, cyclic burrowing or emergence is probably an adaptation that prevents desiccation, or, in other cases, inundation.

Finally, as will be discussed subsequently (cf. section on navigation), rhythms are important in the determination of direction during migration (allowing the possibility of correcting for the apparent movement of the sun), as well as in the timing of migration.

CYCLES IN TERMS OF CONTROLLING STIMULI

Many cycles quickly disappear when constant environmental conditions are imposed on their possessor. Thus, changes in the heart rate of a frog, which may show a regular diurnal alteration, will cease changing when the animal is confined in a constant-temperature chamber. The other extreme is represented by cycles that persist for a considerable time even after isolation from all pertinent and cyclic environmental stimuli, such as the activity rhythms of many rodents. An intermediate situation is afforded by functions requiring an environmental cue for their initiation or timing, but which then run independently of environmental cues. As is to be expected from biologists, many different classificatory and nomenclatural schemes have been promoted to deal with these different kinds of cycles. The more important of these have been summarized by Cloudsley-Thompson (1961), from whose work Table 8–2 has been adapted.

The distinctions between the categories listed are far from absolute. Animals cannot be simply classified as to the nature of their rhythms since in one animal there may exist many cycles, differing in function, phase, frequency, and degree of dependence on environmental timers.

When the cycles are driven by external events, the analysis of their control mechanisms generally poses little problem. One of the earliest cases, Rowan's (1925) investigations of the seasonally cyclic recurrence of reproductive behavior in certain temperate-zone birds and mammals (see review by Bissonette, 1933; Wolfson, 1959; also cf. Chap. 5) showed photostimulation of a certain magnitude to be the relevant timer, e.g., when day length attains a certain value, reproductive activities commence. Recent work by Farner (1959), Marshall (1955), Wolfson (1959), and others has elucidated the pathways along which the reactions initiated by photostimulation must travel, as well as the fact that factors other than photostimula-

TABLE 8–2

AUTHOR	RHYTHMS THAT DO NOT PERSIST UNDER CONSTANT CONDITIONS: BEING INDUCED AND CONTROLLED BY PERIODIC FACTORS IN THE ENVIRONMENT	ENVIRONMENTALLY INDUCED RHYTHMS THAT PERSIST UNDER CONSTANT CONDITIONS FOR AT LEAST A SHORT PERIOD OF TIME	RHYTHMS THAT PERSIST FOR SUBSTANTIAL PERIODS OF TIME UNDER CONSTANT CONDITIONS, THOUGH THEY MAY BE MODIFIED BY THE ENVIRONMENT
Welsh (1938)	Extrinsic rhythms		Persistent rhythms
Park (1949)	Exogenous rhythms	Habitual endogenous rhythms	Endogenous rhythms
Stephens (1957)	Environment-dependent frequencies	Environment-induced frequencies	Environment-independent frequencies
Pittendrigh (1958)	Field rhythms	Impressed rhythms	Endogenous self-sustaining oscillations

tion may be involved. The important matter for us here is the demonstration that the seasonal variation in day length can drive the reproductive cycle. Under constant conditions the rhythm may break down.

When the cycles are largely endogenous, the situation can become more complex. The attempts to explain the persistence of rhythms in animals held under seemingly constant conditions have fallen into two categories. On one hand, one assumes that truly constant conditions are unattainable, that some residual and cyclic variables always remain to impinge upon the isolate, e.g., changes in cosmic radiation (cf. Brown, 1959). On the other hand, we have the assumption that physical or chemical reactions at a subcellular level may oscillate independently of environmental factors and serve as the basic driving mechanism for a biological clock. Such self-sustaining oscillators may be entrained by environmental rhythms whose frequencies do not differ too radically from the endogenous frequencies. This would account for the phase-constancy of most biological rhythms, while the same rhythms, under constant conditions, drift out of phase with environmental rhythms (Pittendrigh, 1958). Other advantages of the second and corresponding weaknesses of the first view, which are beyond the scope of this chapter, are discussed by Cloudsley-Thompson (1961, Chap. 7).

Finally, we must make at least a passing reference to the mechanism of the *clock*. Work in this area is of a rather recent nature, depending as it does on refined neurological, endocrinological, and ultimately biochemical analyses. Whether a classification of rhythms on the basis of their driving mechanism can ever be attained or whether the mechanisms underlying all rhythms are basically identical is as yet impossible to say (cf. Bünning, 1963).

Animal Migration and Orientation

Studies of seasonal movements of animals must be concerned with two distinct aspects: the causative factors *per se* and the processes that allow the maintenance of a particular course. Analysis, in turn, must consider two kinds of causes, distinguished by Lack (1954) (and Aristotle) as *proximate* and *ultimate*. By ultimate are meant the historical or selective factors favoring the evolution of migratory behavior. Those most commonly cited include climatic changes necessitating annual movements (Landsborough-Thompson, 1926), continental drift (Wolfson, 1948), and adaptations for the exploitation of a temporary or fugitive food supply (MacArthur, 1959). A discussion of these explanations of migration is outside the scope of this chapter. (For a review, see Dorst, 1962.) Proximate factors are those which immediately trigger the migratory response. These have been alluded to in the previous section where the role of photostimulation in the induction of breeding behavior was considered (see also Chap. 9). Most of the same comments apply to the initiation of migration (also, Farner, 1959).

The maintenance of a particular direction represents the second distinct aspect of the problem, one which, as we will see, involves an intimate relation between migrations, tropisms, and rhythms. The earliest studies were greatly limited by the inability of investigators to keep track of specific individuals. The actual goals and pathways of migrants could only be indirectly inferred. Marking of individual birds to allow tracing them had been proposed at the start of the nineteenth century by J. A. Naumann (cited by Stresemann, 1951) and again by Borggreve in 1884. Not until 1890 did such a scheme become adopted however; in that year a Danish school teacher, H. C. Mortensen, introduced leg rings (for a history of bird banding, see Stresemann, 1951). Mortensen's bird bands were quickly adopted by ornithologists the world over, and clubs were developed to store and exchange information on recaptures. Thus, a fairly complete picture of migratory pathways of birds was painted, opening the possibility for the experimental study of orienting mechanisms.

One of the earliest and, for its time, most sophisticated of these studies was that of Exner (1893). In order to determine if homing pigeans receive information on the direction of displacement while being transported from home, he moved several birds while they were under anesthesia. Others were transported in swinging or rotating cages (presumably to eliminate inertial clues), while yet others were subject to the discharge from a faradic cell (to interfere with electromagnetic clues). Regrettably, Exner's sample was too small, the response variance too great to allow any conclusions to be drawn. Surprisingly, it took over 50 years for biologists to repeat some of his important experiments!

However, other results were reported that suggested that orientation depended either on an inertial sense or the perception of electromagnetic fields. Thauzies (1898) cited cases in which homing pigeons were delayed by electrical storms (cited by Warner, 1931, as are most of the following references) and Michel (1928) reported that birds in the vicinity of radio stations were disoriented. More recently, similar reports of disorientation as the consequence of radar beams (Knorr, 1954) have been made. Viquier (1882) attempted to explain how electromagnetic sensitivity could provide clues for orientation, suggestions incorporated by Yeagley (1951) in his empirical studies. Yeagley's claims that interference with the electromagnetic sense of pigeons (by attaching magnets to them) prevented orientation were not at first substantiated, though often repeated. Until 1970, it was, in fact, believed unlikely that there are physiological sensitivities of low enough threshold to allow detection of changes in electromagnetic fields (Griffin, 1952).

Exner's other suggestion that orientation involved retracing the directions followed on the outward journey was taken up by Reynaud (1900). However, that retracement is not always involved is known from the direct observations of homing terns made by Griffin (1952).

Not surprisingly, another of the early explanations of homeward orientation involved the assumption that the goal (home) itself provides a beacon that can be seen or otherwise sensed (Hachet-Souplet, 1909). The important arguments of Watson and Lashley (1915) demonstrated that in birds visual mechanisms were simply insufficient to account for long-distance homing, e.g., because of the earth's curvature, light from the top of a 500-mile distant lighthouse would be visible only from an altitude of 25 miles. The suggestions by Wojtusiak (1946) that homing might be accomplished by the sensing of infrared rather than visible light are completely lacking in experimental support, as well as inconsistent with Watson and Lashley's claims (1915). Watson (1915) was unable to demonstrate infrared sensitivity in his studies of the pigeons' spectral sensitivity.

Griffin, in his 1944 review, listed the major theories accounting for orientation and the work that supported them. As with the earlier work cited above (cf. Warner, 1931), these theories involved explanations based upon either vision, kinesthesia, or electromagnetic receptors. Griffin concluded that none of the experimental data favored these theories and proposed instead orientation based on recognition of major geographical or topographical features, typically prevailing air masses, and the relationship between such features of localities near their home and the homeward direction. Griffin also suggested that celestial clues such as the direction of sunset relative to home might be used. This was portentous. Only a few years later, Kramer (1952) and his students were able to demonstrate the existence of a *sun compass* in certain birds (see below). The hectic course of events during this period (1930–1950) has been summarized by Matthews (1955), who also distinguished orientation theories according to rather different criteria than those of Griffin (in 1944).

Matthew's first category included theories that required the maintenance of sensory contact with "home." It was theories in this category (and the second, below) that were rendered untenable by Griffin's (1944) criticisms. The second category included theories of navigation "by means of a 'grid' derived from the earth's rotation and magnetism" (Matthews, 1955, Chap. 7), and the third, a grid derived from the sun's co-ordinates. To this last, one would today have to add, "or co-ordinates of other celestial bodies" (e.g., Sauer, 1957). It is theories in this last class that have proven most viable (see below).

After 1950, the field of orientation made advances as great or greater than those made in any of the other ethologically related schools. Before continuing, it is useful to note certain stated distinctions between the different kinds of orientation mechanisms. The three categories we shall distinguish have been characterized by Schmidt-Koenig (1965: 218–219) as follows:

Piloting—to find a goal with reference to familiar landmarks. Upon displacement, animals (a) either execute a random search, or (b) perform a systematic search for relevant landmarks.

Directional orientation or compass orientation—to head in a geographical direction without reference to landmarks. Goal orientation with the capacity of compass orientation is only possible if taking up a certain direction always leads to the goal or familiar landmarks so that piloting may then be employed. Terns (*Sterna hirundo*) in coastal New England, for example, fly southeast when displaced inland. This direction brings them to the coast which they can follow until familiar landmarks are encountered.

Navigation—the capacity to maintain or establish reference to a goal other than through recognition of landmarks (the final approach to the goal being excluded). Many birds must be able to navigate, as indicated by many phenomena of migration, and as demonstrated in displacement experiments. Two major subclassifications are necessary:

(a) *Reverse displacement navigation—to reach a goal by reversing the actual or summed displacement through information gathered during the displacement process.* This subclass may include several different, but similar, kinds of capacities. For instance, an animal with compass orientation capacities plus a method for measuring distances and a memory and/or an integrating mechanism could either retrace its steps during displacement or could sum the displacement and compute the direction and possibly distance of the goal. The sun compass is used for keeping track of the direction(s) of the outward journey and for steering the computed homeward course. The limiting factor is the availability of the sun. Another mechanism for reversing displacement would be a sensory system which continuously summed changes from the "home" setting, such as in inertial navigation systems. In a "one-step" operation, the displaced animal need only to take up a heading which diminishes the precession or change from the home value. "Hybrid" inertial navigation systems can be imagined which utilize a compass for the actual homing once the inertially obtained information of direction of home has been transformed into the modality of the compass. Such a hybrid system would represent a "two-step" operation. Pure or hybrid inertial navigation systems have not been actually demonstrated in animals.

(b) *Bi-coordinate navigation—to establish reference to a goal by comparing two (or more) coordinates (astronomical, geophysical, etc.) of the displaced position with the quantities of the goal position, without utilizing information gathered during displacement.* This subclass may also include several different, but similar, kinds of capacities. Such systems would represent "one-step" operations in which the animal just moves in such a direction as to "reset" the coordinate quantities to the goal values. Again, "hybrid" systems could be thought of that utilize a compass in a "two-step" fashion in analogy to hybrid inertial navigation systems. Pure or hybrid bicoordinate navigation systems have not actually been demonstrated in animals.

Piloting is of little direct concern to us here, although the possibility that systematic search patterns may be adopted by birds seeking their home raises several interesting problems. Directional orientation on the other hand makes other demands. Although it is possible to conceive of simple tropisms as providing the basis for directional orientation, Kramer and his group (1952) have demonstrated that for many birds (as well as other animals, e.g., reptiles, Fischer, 1960; insects, Renner, 1960, and Lindauer. 1961) it is the position of the sun that is of prime importance. Kramer (1952) was able to train his starlings to seek food from a concealed aperture lying in a particular compass direction along the periphery of a circular cage whose opaque sides excluded sight of all landmarks. Only sky and sun were visible to the birds. At whatever time of day they were tested, they were able to maintain the training direction. When the apparent position of the sun was shifted by means of mirrors, the choices made by the birds shifted a corresponding number of degrees. Hoffman (1953) then subjected trained starlings to artificial day-night cycles which began either six hours earlier or later than the natural cycles. His birds, when tested under the natural sun, also shifted their choice points 90 degrees to the right or left, respectively. These experiments have since been repeated with a number of other species (cf. review by Schmidt-Koenig, 1965) and similar experiments performed under the night sky with only stars visible (Sauer, 1957). Thus, it is clear that these birds are able to keep track of the passage of time and to compensate, when orienting, for the movement of the sun (or certain stars) across the sky. This *sun compass*, which allows maintenance of a fixed direction, is thus seen to be intimately dependent on a 24-hour internal clock or endogenous rhythm.

Navigation, as Schmidt-Koenig (1965) points out, may involve a sun-azimuth compass and chronometer, but in addition certain other components are needed. These latter must include either mechanisms for recording the direction and distance of the displacement from home or an "almanac" that allows the animal to compare the co-ordinate of its home with its displaced position.

An interesting sidelight that relates the clockwork to the neuroendocrine system was revealed in a study by S. Emlen (1969). He found that a particular star pattern (simulated in a planetarium) could trigger directional movements either southward or northward, depending on whether the birds had previously been subjected to a long-day (springtime) or short-day (autumn) regimen.

Any of a number of inertial navigation systems would meet the necessary requirements. Barlow (1964) has detailed the characteristics of some inertial devices and discussed the possibility of there existing biological analogues. Thus far, there is no compelling experimental evidence either for or against inertial theories. The possibility that inertial and other, e.g., visual, mechanisms co-exist in one animal and can replace one another (hybrid systems) depending on environmental conditions makes the design of critical experiments exceedingly complex.

As for visual orientation systems, these all appear to depend on the existence of a grid. One of the two necessary co-ordinates of such a grid appears to be based upon the constancy of the sun azimuth at a particular locality, date, and time. Knowing what the sun azimuth should be at noon on a particular day at one's home, for example, one can readily determine whether one is actually at home or has been displaced to the east or west. In theory, information on displacement to the south or north should be provided by comparable data on sun altitude. This, in fact, is the basis of Matthews' (1955) sun-arc hypothesis (Fig. 8–1). Modifications of this have also been offered (Pennycuik, 1960), whose grids were based upon values for local rates of change in altitude and azimuth. In laboratory and field experiments, however, the altitude of both an artificial and the real sun have not been seen to influence directional choices of birds (Schmidt-Koenig, 1965), although fish may make some use of information on solar altitude (Braemer and Schwassmann, 1963).

Unfortunately for aspiring theorists, a number of other seemingly contradictory results had yet to be reconciled. While, on the one hand, Kramer and his students had repeatedly noted disorientation of birds tested under heavily clouded skies, i.e., with no sun visible, Gerdes (1962) and Fromme (1961) had reported orientation in the absence of all celestial clues. Fromme's old world warblers were tested in a circular cage, fitted with perches about its periphery that automatically recorded the arrival or departure of a bird. During the spring and fall, captive migrants frequently become very restless, flying to and fro and orienting themselves along their usual migration track. Fromme's birds maintained this orientation both inside an opaque shelter and the confines of a climatic chamber. Only when the steel door of this chamber was closed did the directions chosen by the birds appear to scatter randomly. It is possible that this and similar results are artifacts of small samples and inadequate or inappropriate statistical

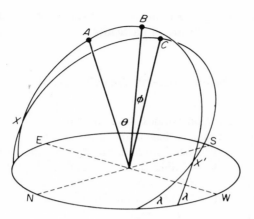

Fig. 8–1. Schematic diagram of the sun-arc hypothesis. If the observer is standing on the surface of the earth at the point in the center of the diagram, he observes the sun to travel the arc λ-X'-C-X. At noon (home time) the sun is at the highest pont of its arc, C.

Now, if the observer is displaced to the southwest, and the arc of the sun from the new position is graphed in relation to the observer in the center of the diagram, the new arc is λ-X'-B-A-X. At noon (home time) the sun is at A (not at B, the highest point in its arc).

Matthews hypothesizes that the observing bird extrapolates from the movement of the sun at A, the theoretical highest point of the arc B. The difference between A and B (i.e., the angle θ) is a measure of the westward displacement from home, while the difference between B and C (the angle φ) is a measure of the southward displacement from home.

In any case, if the sun is lower in the sky during the morning than it should be according to the bird's internal chronometer, the bird is west of home; if the sun is higher, the bird is east of home. During the afternoon the relations are reversed. North-south displacement is more difficult to explain, for in Matthews' theory it requires the bird to extrapolate the sun's arc from a short period of observation of its movement. Pennycuik's (1960) modification would have the bird measure the rate of change in altitude of the sun, probably an easier judgment to make in terms of sensory mechanisms. In either theory, north-south displacement cannot be judged instantaneously—as can east-west displacement—but depends on some measurement of the sun's actual movements.

treatment. However, at the present these results remain somewhat disconcerting to proponents of exclusively celestial theories (but note Recent Developments below).

One other experiment must be mentioned for the support it gives proponents of inertial theories: Migratory pigeons, after having been rotated, show a peculiar pattern of electrical activity in the cerebellum. Nonmigratory races fail to show the same pattern (Gualitierotti et al., 1959). It is because the cerebellum is so intimately involved with balance and co-ordination that these results are of particular interest and importance.

Some Recent Developments

Since 1970, orientation studies have increasingly focused on the non-visual modalities. Animals other than birds, arthropods, marine turtles, and amphibians have also been studied, but most of the effort continued to be focused on the avian navigational system, and particularly that of pigeons (see reviews by Griffin, 1969, and Galler et al., 1972).

In the 1960's, studies by Merkel and by Wiltschke re-opened the question of magnetic sensitivity. Yeagley's (1947) findings of interference effects from magnets had previously been rejected by most biologists, so these newer studies were initially treated skeptically. By 1971, when Keeton asserted, "Magnets interfere with pigeon homing," the skepticism began to be replaced with new experiments. Experiments in which pigeons were fitted with frosted semi-opaque contact lenses but still homed (Schmidt-Koenig and Schlichte, 1972) also indicated that the eyes, while normally used, need not play an essential role in homing. Further experiments utilizing birds wearing magnetically opaque hoods or birds trained to discriminate magnetic fields in a Skinner box are underway even as these lines are being written. An assessment of the information available in the earth's magnetic field lends support to the optimism with which these studies are being pursued (Freedman, 1973).

Overview

The directed movements of plants and animals have been the concern of biologists since the start of the nineteenth century. Animal tropisms, in fact, provided the basis for the mechanistic schools of behavior that came to the fore at the close of that century. Loeb argued that all animal movements could be understood in terms of *tropisms*. Specifically, he believed that whenever a source of stimulation impinged asymmetrically upon the sense organs of a bilaterally symmetrical creature, a similar inequality in the magnitude of muscle contraction on opposite sides would result. This view was attacked by Mast and later by Fraenkel and Gunn, who were able to demonstrate a variety of rather different processes underlying the orientating responses of different animals.

Studies of biological rhythms go back a far shorter span of time. The first important ones can scarcely be traced beyond Forel and von Buttel-Reepen at the turn of this century. Subsequently, various workers concerned themselves with such aspects of cycles as their frequency, overt effects, and adaptive functions. During the present decade we have come to recognize a wide variety of rhythms, ranging from those seemingly independent of external timers to those which vary directly with environmental conditions. A detailed understanding of their underlying mechanism has yet to be reached.

The study of animal navigation received its major impetus from the invention of bird bands, a technique that allowed the establishment of migration routes. As early as 1893 experiments were undertaken to determine the guiding cues in migration, including visual, electromagnetic, and inertial stimuli. It was much more recently, however, that the role of the sun and of internal timers in navigation became recognized. Kramer's discovery of the avian sun compass was the major step that has added so much to our understanding of navigation and has permanently wedded the fields of biological chronometry and navigation.

Selected Readings

The best review of directional movements still is G. S. Fraenkel and D. L. Gunn's volume, *The Orientation of Animals* (1940), recently republished by Dover Press. The classical reviews on bird navigation are by G. V. Matthews, *Bird Navigation* (1955), or Carthy (1956), while more recent coverage is provided by K. Schmidt-Koenig's review (1965). The migration of birds, a subject not explicitly developed in this chapter, is well-covered in a volume by J. Dorst (1962). Radar observations of migrating birds and correlations with meteorological conditions are to be found in Lack (1963). For an excellent selection of papers on both cycles and navigation, refer to the *Cold Spring Harbor Symposium on Quantitative Biology*, Vol. 25 (1960); for the treatment of cycles alone, see Bünning's book (1963) and the *Annals of the N.Y. Academy of Sciences*, Vol. 98 (1962): 753–1326.

Part IV | ETHOLOGY
TODAY

Introduction

From the concluding sections on recent developments in the chapters of Parts II and III, it is clear that ethology is expanding at a tremendous rate. Can today's studies tell us what ethology's next century will be like? Probably they cannot provide specific details, although two points can be prophesied.

First, ethology will have a growing influence on other fields of study. Details of such influences lie outside our purview, but some trends are clear. The application of behavioral characteristics in classifying animals and reconstructing animal phylogenies already has a firm beginning (e.g., Simpson and Roe, 1958). Ethological results will continue to influence psychological and psychoanalytic thinking (e.g., Ploog, 1964). Finally, ecology is turning increasingly to behavioral analyses of some important problems (e.g., Klopfer, 1973).

Another extrapolation from the present concerns the general domain of "explanations" of behavior. Although we cannot predict what specific problems will arise tomorrow, ten years hence, or in the next century, we can say that they will be parts of a framework of ethology recently made explicit by Tinbergen (1963) and Hailman (1964a). That is, each problem will ultimately contribute to our knowledge of one or more of the "causes and origins" of behavior (1) How is behavior maintained in a population? (2) What is the history of behavior in a population? (3) How is behavior controlled? (4) How does behavior develop ontogenetically? In the following chapters we can cite only a few examples of the exciting recent studies that contribute to these questions. (The concluding sections of the chapters in Parts II and III are also germane but will not be unnecessarily repeated here.)

Chapter 9

HOW IS BEHAVIOR MAINTAINED IN A POPULATION?

The Biological View of Function

Introduction

Almost all schools of ethology have been concerned in some way with the "function" of behavior; i.e. how does it contribute to the animal's survival or reproductive success? What is the nature of the selective pressure that maintains it? Let us first consider some recent studies demonstrating how behavior is of direct selective advantage and then turn to behavior patterns that are themselves of no advantage but are nonetheless maintained in a population. The Recent Developments sections of Chaps. 2 and 7 also mention current studies dealing with functional aspects of communication and other behavior.

The Selective Advantage of Behavior

At least four approaches to studies of the selective pressures maintaining behavior patterns have been devised. Ethologists owe to Niko Tinbergen of Oxford much of the impetus for differentiating these methods and for showing the importance of studying the survival value of behavior. For a more complete exposition of the problems and methods concerned in these studies, see Tinbergen (1963).

MEASUREMENT OF SELECTIVE PRESSURES

If a behavior pattern varies within a population, it may be possible to study the differential survival and reproductive rates resulting from the various forms of behavior. A good example of the method is Patterson's (1965) study of colonial nesting habits in gulls. With far more suitable nesting habitat available than is actually utilized, why do gulls tend to nest in clumps? In other words, what selective pressures promote the behavior "nesting near other gulls"?

This behavior varies in a population of black-headed gulls. Some gulls nest very near other gulls (i.e., those in the center of a colony), while others nest less close (i.e., those at the periphery). Patterson found that those nests near the periphery are more often robbed by predators, which eat the eggs or chicks. Thus, those gulls with the strongest tendency to nest with other gulls (i.e., those in the center) leave a higher proportion of offspring. These offspring, being genetically similar to the parent, will thus show a high incidence of social nesting in the next generation, when their young will have the highest probability of survival. Thus, the social nesting behavior will be maintained in the population.

Another example of selective pressures acting on behavior is provided by Roeder and Treat (1961). Some moths, upon detection of the ultrasonic cries of bats (cf. Chap. 6), execute various forms of escape behavior; they may fly away from the sound source, fly erratically, or plummet into the grass and hide. Not all moths have tympanic organs with which to detect ultrasounds of bats, and it is not certain that all moths with the organs always execute escape maneuvers when detecting the sounds. Nevertheless, it was possible to measure the overall selective advantage of the detection-*cum*-evasion behavior in moths.

By filming an area lighted by spotlights, Roeder and Treat (1961) noted the differential survival of moths that did and did not execute evasion upon attack by a bat. By counting the outcome of 402 such encounters, they found that for every 100 reacting moths that survived, there were only 60 surviving nonreactors. They call the difference between these figures a selective advantage of 40%.

Of course, the selective advantage possessed by the evading moths is at least partially dependent on their not constituting too large a majority of all moths. As in the case of mimetic resemblance, where a palatable butterfly escapes predation because of an appearance similar to that of a distasteful model, the success of the ploy depends on the predator not becoming "smarter." If the palatable mimic becomes commoner than its model, the predators may not mind paying the price of an occasional unpleasant mouthful. Similarly, when most moths perform an evasive maneuver, the bats may be expected to evolve a countermeasure, for instance, ultrasounds of a frequency beyond the moth's rage. The advantage of the "preferred" behavior pattern, in short, demands variability or polyethism for its expression (see Klopfer, 1973). When everyone has a million dollars, no one's dollar is worth much!

MANIPULATION OF SELECTIVE PRESSURES

Parent gulls remove (or eat) the egg shell and fragments from the nest shortly after the hatching of their chick. Why should such a seemingly inconsequential behavior pattern, occupying only a few moments per year in the life of a gull, be fastidiously performed? By constructing a simple naturalistic experiment, Tinbergen and his co-workers (1962) showed that the probable selective pressure is provided by predation.

Since the shells are invariably removed by the parents, there is no natural variation such as exists in social nesting (above). Therefore, nests (without parents, of course) were placed near a gull colony, and in these nests were placed various combinations of eggs, chicks, and egg shells. The eggs and chicks of *ridibundus* are colored with splotches of brown, green, and khaki, but the interior of the egg shell is, of course, white. Consequently, crows and other predators more rapidly and more often discovered the unguarded experimental nests with egg shells than they did the similar nests lacking the shells. Parents that remove the egg shells clearly would leave more offspring than those not removing the shells.

INFERENCE FROM COMPARATIVE STUDIES

Cullen (1957) found that the cliff-nesting kittiwake differs from ground-nesting gulls in many respects: it has different display signals; it fights in a special way; it does not remove the egg shells; it builds its nest out of different materials and with different movements, and so on. She hypothesized that all these differences are due to selective pressures (from reduced nesting space) that accompany cliff nesting but do not accompany ground nesting of ordinary gull species. Since no direct measures of survival and reproduction were made, Cullen's (1957) observations constitute only an hypothesis, not proof, of the forces of natural selection. But her observations do serve as a prediction of the nature of behavioral characteristics to be found in other cliff-nesting gulls.

Fortunately, another cliff-nesting species, heretofore unstudied, was available to test the hypothesis—the swallow-tailed gull of the Galapagos Islands. Hailman (1965) noted in this gull the environmental conditions (i.e., the presumed selective pressures) and the morphological and behavioral characteristics that Cullen (1957) considered correlated with cliff nesting. In adaptations to reduced living space and the danger of eggs' or chicks' falling over the nesting cliff, the swallow-tailed gull shared eleven characteristics with the kittiwake, six with ground-nesting gulls and one characteristic was intermediate. When the swallow-tail's environmental conditions were intermediate between those of the kittiwake and ground-nesting species (scarcity of nest sites, degree of nest predation), the adaptations to these conditions were dissimilar to those of the kittiwake: five were shared with ground-nesters and two of the characteristics were intermediate. Finally, when conditions resembled those of the ground-nesters completely (scarcity of nesting materials), the swallow-tail shared all five of the presumably correlated characteristics with ground-nesting gulls only.

This high correlation between factors of the environment and the predicted behavioral characteristics demonstrates the utility of comparative studies performed in this way. Even the exceptions (environment similar to the kittiwake's but characters resembling those of ground-nesters) can probably be attributed to quantitative differences between environments and characteristics. The important aspect of this method is that predictions made from a comparison of two or more species can be confirmed (or rejected) by examination of another previously unstudied species. The obvious drawbacks to this approach are considered in Chap. 10 in which we turn more explicitly to the issue of arguments from analogy.

MODELS OF NATURAL SELECTION

Yet another way of studying the effects of natural selection in maintaining behavior is to set up a population model. For instance, it would be possible to create an environment in which the usually useful behavior pattern of flying would be disadvantageous to fruit flies. If a strong fan were blown through the cage of a population of *Drosophila*, most of the animals that tried to fly around might be thrown against a wall to perish. Mutants with vestigial wings or with less motivation to fly, who stayed on the bottom of the cage, could feed and multiply. Eventually, only nonflying flies should comprise the population.

Other Modes of Maintenance

Phenotypic characters may be maintained that are not themselves of direct selective advantage to their possessor (e.g., R. A. Fisher, 1930). Some of the processes responsible for this phenomenon may operate on behavioral as well as morphological characters (for instance, the lack of selection in small populations and pleiotropism).

ABSENCE OF SELECTION IN SMALL POPULATIONS

If a population of animals is very small, there may not be enough behavioral variation upon which natural selection can work. In other words, the behavioral constitution of the population may remain static (and possibly not fully adaptive) in the absence of any natural selection. Curio (1961) discovered that the Spanish mountain population of the European pied flycatcher, *Ficedula hypoleuca iberiae*, differs from northern populations of the same species, *E. h. hypoleuca*, in various details of seventeen behavior patterns (e.g., song, pairing, displays, etc.), even though the environments in which they live are essentially identical. It seems, then, that these traits in the small Spanish population are maintained because there is too little individual variation upon which selection can work; otherwise, natural selection would presumably adapt each of the populations in the same way. Only the absence of mobbing responses to a predator, the red-backed shrike, appears to be adaptive, since the shrike does not occur in Spain (Curio, 1961). The question of their differences then becomes one of historical origin and will be considered more fully later (Chap. 10).

PLEIOTROPISM

If a neutral or disadvantageous character is maintained in a population because of genetic correlation with another character which is being selectively favored, the phenomenon is called *pleiotropism*. A behavioral example has been suggested by Hailman (1964b).

The Galapagos swallow-tailed gull shows a high synchrony of nesting activities within local groups. Darling (1938) has suggested that the capacity to respond synchronously was specifically favored. However, the advantages proposed by Darling apply only to the whole colony, and whole colonies of the swallow-tail are not synchronous. Although any advantages of colonial nesting in general (see Chap. 7) depend on more or less synchronous breeding, the exact synchrony of local groups is not fully explained by any specific advantages.

Hailman (1964b) suggests that local synchrony is merely a manifestation of normal breeding behavior. Since the members of a pair must respond to one another's displays through the breeding cycle, it seems likely they will also respond somewhat to the same displays of neighbors. Thus, pair A, being advanced in the breeding cycle, will accelerate the physiological and behavioral breeding cycle of neighboring pair B; likewise, pair B's displays may retard the cycle of pair A somewhat. The result of these incidental responses to neighbor's displays is, of course, local breeding synchrony.

OTHER POSSIBLE FACTORS

The previous discussion has assumed that behavior is under direct genetic control and can, therefore, be treated like any morphological character. As will be mentioned in Chap. 12, this is a gross simplification. But, whatever the underlying mechanisms of any particular behavior pattern, the patterns themselves are ultimately subject to selection—at least in the natural situation.

It should also be possible for a behavior pattern of negative selective value to be maintained in a population if a genetic mutation rate producing this pattern is greater than the selective pressures against it. We know of no natural situations in which this has been shown to be true for a behavior pattern. However, if the individually disadvantageous behavior represents *altruism*, a special case is presented.

By altruism we mean that the behavior in question reduces the probability that its practitioner will survive to leave offspring even while enhancing the prospects for another animal (usually a conspecific). Wynne-Edwards (1962) has argued that "group selection" accounts for the maintenance of altruistic behavior, whereby he presumably means that advantages gained by a group may so raise the fitness of that group as to compensate for the decrease in fitness suffered by individuals. ("Fitness" means the proportionate contribution of an individual or group to a future gene pool. The larger the proportion of the genes in any generation that can be attributed to a forebearer, the greater that forebearers' fitness.)

There is no evidence (Williams, 1971) to show that group selection actually operates, although Hamilton (1963) has shown that altruism can evolve if the benefits are proportional to the closeness of the genetic relationship between altruist and recipient. Hamilton has shown that if an altruist leaves no offspring because of his act, his altruism may yet be favored if he enhances the fitness of at least two of his siblings or offspring (since each of these will have 50% of their genes in common with the altruist). Thus, the altruist's "fitness" is the same whether he lives and two of his brothers die or *vice versa*.

In this restricted sense of group selection within families, behavioral examples can easily be found among distraction displays. A killdeer caught by a predator during the "broken wing" display may fail to survive, but it may also ensure the survival of four young offspring of similar genetic constitution.

Finally, it does seem likely that inappropriate behavior patterns that are largely learned through social traditions or early imprinting-like processes could still persist in a population, at least until the population becomes extinct. For instance, Heusser (1960) has shown that the European toad continues to return for breeding to the place in which it grew up, even though the suitable breeding habitat of this place has been completely obliterated by man. This results from the rate of changes in the environment being of a different scale (due to human intervention) than the rates at which natural selection operates. Traditional migratory routes of waterfowl may present a similar case (Hochbaum, 1955). However, such maintenance is transitory, for the population either does alter its migratory behavior or perishes.

These discussions of the maintenance of a behavior pattern in a population are not meant to be a final word. There may certainly be other and more complex factors at work than those mentioned here. But these studies give some idea of one branch of behavioral research that will continue to command the attention of ethologists in the future.

Selected Readings

Klopfer, P. H. 1973. *Behavioral Aspects of Ecology*, 2nd ed. Prentice-Hall, Englewood Cliffs, N.J. (This book provides many additional examples of the effects and action of selection.)

Kruuk, H. 1964. Predators and anti-predator behavior in the Black-headed Gull (*Larus ridibundus* L.), *Behaviour supplement 11*, 129 pp. (A monograph showing many examples of the selective pressures of predation by foxes, crows and other animals upon the behavior patterns of a colonial bird species.)

Tinbergen, N. 1964. Behavior and natural selection. *Proc. XVI Int. Congr. Zool. 6* (A plenary symposium talk delivered at Washington, D.C. in August, 1963, reviewing studies to that date and demonstrating the need to study natural selection empirically.)

Chapter 10

WHAT IS THE HISTORY OF BEHAVIOR IN A POPULATION?

Generation Changes in Behavior

Introduction

There occur in the woodlands of Eastern North America three super-ficially similar ground-dwelling birds: a sparrow, a thrush, and a mimic-thrush. All feed on insects and grubs hidden beneath the leaf litter, yet each turns leaves in a different manner. The fox sparrow kicks the leaves back violently with a two-footed movement resembling a hop. The wood thrush extends one foot and quivers it in the leaves, stirring a small circle. The brown thrasher stirs with its bill, often grasping single leaves and throwing them aside. Since each of these three behavior patterns is maintained in the species through essentially the same selective pressure (feeding efficiency), why do they differ? That is, how can the exact form of each of these be-havior patterns be explained? Like the apparently random house numbers on the streets of Oxford, there can only be a historical explanation.

At some time in the distant past a sparrow, a thrush, and a mimic thrush began feeding in leaf-litter. They presumably evolved different be-havior patterns for doing this because the relative ease with which par-ticular patterns could be evolved differed among the species. For instance, it may have been relatively easy for a sparrow to kick back with the legs, rather than to push forward as in a normal hopping movement; those in-dividuals so responding survived and reproduced at higher rates than those that did not. Perhaps it was easy for a thrush with an upright posture to extend one foot forward, since thrushes often run and walk (that is, use the legs independently) rather than hop. Perhaps also it was easy for the ground-dwelling thrasher with a long, horizontally held body to use its bill instead of its legs. Thus, each species may have been anatomically predis-posed in some way to evolve one of the leaf-turning patterns instead of another. On the other hand, perhaps the anatomic differences are trivial.

This example demonstrates the need to consider the historical origin of a behavior pattern in a population in order to understand the pattern fully. However, historical studies must be largely speculative since there are no behavioral fossils, *per se*. Having no long historical record of the develop-ment of behavior, we must depend on inferences from comparative studies or on very short historical records from which we extrapolate lengthier behavioral histories. These severe limitations of method demonstrate why very little is known of the history of behavior. Some recent attempts at the reconstruction of behavioral phylogenies are mentioned in Chap. 2.

Natural Selection through Time

The most obvious historical cause of a behavior pattern is the action of natural selection upon variable populations. Here the study of the history of behavior converges with problems of maintenance that were discussed in the previous chapter. However, there do exist several possible methods for studying the phylogenetic history of behavior.

BEHAVIORAL "FOSSILS"

The structures underlying behavior or the results of behavior can sometimes become fossilized. For instance, bony structures may reveal whether or not a particular fossil animal could fly; footprints of extinct reptiles allow reconstruction of their gait; fossil termite nests reveal something about nest-building behavior; and insects imbedded in amber may reveal sociality. Other examples are cited by Wickler (1961). To establish direct relevance to a behavior pattern in an extant animal, however, one must establish that the creatures leaving the fossilized remains are actually ancestral to the animal under study.

TRACING SELECTION IN THE FIELD

When some particularly great change in the environment of a rapidly reproducing species occurs, it may be possible to trace changes in the composition of the population due to natural selection (also cultural changes, see below). For instance, Kettlewell (1965) has shown that the morphology of some lepidopterans in England has been altered by industrialization. The increased soot put into the air has been deposited upon trees, making the moth's substrate much darker than it was in former times, and much darker than it is in nonindustrialized areas today. In industrial zones, the darker moths of the population have more readily escaped predation and reproduced at a higher rate, whereas the lighter phases of the moths predominate in nonindustrialized regions (and in collections from the now-industrialized regions that were made prior to the industrial revolution). It has been suggested that correlated with this morphological change is a behavioral change; in industrial regions, moths rest preferentially on dark backgrounds. As anti-pollution ordinances take effect and soot disappears, the incidence of melanism should decrease, along with the preference for dark background.

SELECTION DURING DOMESTICATION

In domesticated animals we find the end point of an artificial selection imposed by man that is essentially a special case of natural selection. Actually, there are two sorts of selection that may take place during domestication. The first involves conscious selection and selective breeding for specific characters, e.g., barklessness in dogs, docility of various species, or egg productivity in fowl, without respect to the effects of such selection upon general fitness. Such selection does not truly mimic natural selection since the overall fitness of the animal is not at stake; *natural* selection works on whole organisms, not individual characters.

The second sort of selection involves the more subtle action of the man-created domestic environment. Herre (1955) has listed a number of changes in brain morphology that are common to the domestic forms of a wide range of species, from dogs to ducks. Apparently, some as yet undefined conditions of domestication affect all domestic stock, irrespective of the particular kinds of characters for which the stock in question is being selected. Thus, the study of the behavior of domestic animals can be especially illuminating, particularly when the wild "ancestor" is still extant.

Phylogenetic Inferences

The ethologist must usually rely on various indirect rather than direct methods for inferring the phylogeny of a behavior pattern. These methods of inference are enumerated below and are followed by a general critique.

RECONSTRUCTING THE ENVIRONMENT OF THE PAST

If one knows how selective pressures from the environment shape behavior patterns and maintain them in populations (see Chap. 9), then it should be possible to deduce something of the history of a pattern by knowing the history of the environment. At present, the environmental history of only a few regions is known at all well and then only grossly. The use of index fossils and the pollen records from limnological studies has made it possible to say something about temperatures and rainfall and to describe the dominant flora of times past. Historical geology has added knowledge of the physical substrate. If, in addition, we were to have fossil ancestors of the recent animal under study and were able to correlate the occurrence of these fossils with particular environments, then it might be possible to extrapolate to the history of the animal's behavior. The severe limitations of this method are obvious; to date, no detailed reconstruction of behavioral history by this method has been attempted. However, one example of this approach is provided by Bartholomew and Birdsell's study of the rainfall patterns of Australia and their influence upon the size of early human communities (1953).

"STRONG" COMPARATIVE STUDY

One method for uncovering the phylogeny of behavior by inference is to find a group of related species that show a graded sequence in some pattern of behavior. This *strong comparative inference* indicates only the steps possibly taken in evolution, not the direction of change. However, if many recent species show a type of behavior at one end of the spectrum and fewer species show behavior at the other end of the spectrum, the former type is generally believed to be primitive (that is, characteristic of the common ancestor). To make this method clear, consider an example from the courtship of "dancing flies," as described by Schneirla (1953).

In some species of these dancing dipterans, the empiid flies, pairing and mating require an elaborate courtship during which the male presents the female with a compact bundle of web that he secretes in flight (see Fig. 10–1). In another species a small plant fragment is encased in the silken web prior to its presentation. In yet another a small fly is so encased, though the actual presence of the prey within the silk capsule is apparently not essential; it frequently falls out during the transfer of the capsule from male to female. Then there are species in which a fly is caught by the male and transferred to the female, who devours this prey during copulation (should human analogies be drawn?). Finally, there are species that merely mate without any presentation of prey or other artifacts. If we know that many more species of empiids show one rather than the other of these courtship patterns, we assume that the more common pattern is the more primitive and thereby provide a basis for inferring a phylogeny.

Another example comes from Wickler's (1962) studies on mouth-breeding cichlid fish. The male possesses spots on the anal fins that resemble the eggs. The spots, also used in courtship, elicit snapping by the female. She snaps at the spots near the male's ventral-posterior side. Thus, she takes sperm into her mouth. She also takes in her eggs. She thereby assures fertilization. A comparative sequence of intermediate forms suggests how this behavioral system may have developed from the ancestral type.

"WEAK" COMPARATIVE STUDY

It is rare that such a convenient series of intermediate behavior patterns occur in related species as is true of dancing flies and mouth-breeding cichlids. More often, each species of a group will have evolved in parallel

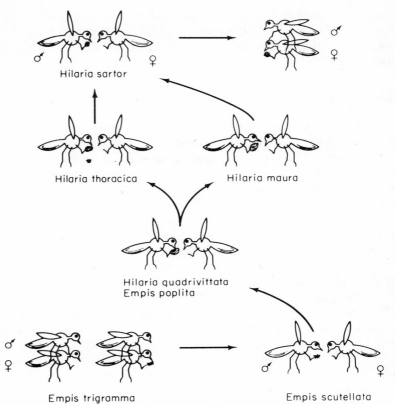

Hilaria sartor

Hilaria thoracica

Hilaria maura

Hilaria quadrivittata
Empis poplita

Empis trigramma

Empis scutellata
Empis borealis

Fig. 10–1. A highly diagrammatic representation of the evolution of courtship in dance flies. The male dance fly, *Hilaria sartor* brings a small ball of secreted silk to the female who plays with it while the male copulates (upper right). How has such a strange pattern evolved? A comparative series of related fly species shows the probable course in the evolution of this behavior. Most flies, like *Empis trigramma*, copulate without elaborate courtship (lower left). However, the voracious female may try to eat the male before or during copulation; if the female is already eating a small prey, she does not attack the male (at least until copulation is complete). In some species (*E. scutellata* and *borealis*) the male catches a prey animal and gives it to the female to eat during copulation. *Empis poplita* secretes a huge balloon of silk around the prey, and *Hilaria quadrivittata* wraps the prey in many silk threads. This wrapped prey seems to keep the female occupied for longer periods than does the prey alone; the female turns the wrapped prey over and over. In *Hilaria thoracica* the wrapping is not firm, and the prey often falls out; in *H. maura* the male wraps any object, such as a petal or leaf. In both cases, however, the female stays occupied with the ball, even though no actual prey is contained therein. Finally, the male *Hilaria sartor* does not waste energy catching a prey or finding a small object, but merely secretes a small ball of silk which is given to the female.

from the ancestral type so that there is little variation among recent species. Or, alternatively, each will have evolved away from the ancestor along its own path of specialization. If the latter be the case, a *weak inference* from comparative study can be made. The inference assumes that the elements of a behavior pattern that are common among the various species represent the ancestral traits.

For example, Tinbergen (e.g., 1953*b*) initiated a study of the communicative behavior of gulls which has been carried on by a number of his students, associates, and others. A "progress report" of this project (Tinbergen, 1959) has shown that the postures, movements, and calls of all species show common elements; it is as if they were all manufactured on the same template, but with large or small variations. The repertoire of a hypothetical ancestor could be made from these common elements, and the forms of divergence from it enumerated (as was done tentatively by Tinbergen, 1959).

INFERENCE FROM VARIATION IN THE INDIVIDUAL'S BEHAVIOR

Finally, inferences about the history of a behavior pattern have been drawn from variations in the individual organism's behavior. Individuals may show gradations of form, intensity, etc., of a single behavior pattern. The spectrum of the individual's variations, like the spectrum of variation in a population, may indicate the evolutionary sequence—or it may not. Such individual variations may be immediately adaptive and have no relation to the phylogenetic or historical sequence; there would seem to be no *a priori* method for distinguishing these possibilities. As in population variation, the direction of evolution here is impossible to establish. There seems to have been no serious application of this approach to establishing phylogenies of behavior.

Another variation involves the development of signal functions or communicative behavior (i.e., displays and similar behavior) from noncommunicative behavior. Examples of how "display" or communicative postures and movements may be derived from simple incipient locomotory movements are given by Daanje (1950). For instance, the tail-flick of a bird taking off from a limb may be standardized to a tail-up posture, or an exaggerated tail-flick movement that acts as a signal to other birds. Both the ordinary flight-preparatory flicks and the exaggerated signal movements may be seen in the same individual. The history of such displays is inferred from their occurrence in noncommunicative contexts.

INFERENCES FROM ONTOGENY

During the development of an individual, its individual structures (and behavior) must also develop. Variations in the heredity of the organism, due to mutation, recombination, or chromosomal aberrations, act on characters by altering them at some stage of development. Since all parts of the organism affect one another in some way (i.e., nearly all genes have pleiotropic effects), a change in one gene may affect the entire developing system. Thus, a genetic change that alters this carefully balanced system early in development is more likely to disturb the organism grossly than a gene change affecting only late stages. For this reason, viable variations in organisms tend to occur late in development, *on the average.*

Natural selection works on variation within a population. Since variation is most often achieved through changes or additions at the end point of a developing character, the ontogeny of the character may, in a vague way, resemble the phylogeny of the character. It is this vague correlation that led Häckel and other early biologists to propose the *biogenic law,* "ontogeny recapitulates phylogeny."

It should readily be appreciated that the amount of information about phylogenetic history that one can derive from the study of ontogenetic sequences is both minute and imprecise. It is possible to change the ontogeny at any stage, but it is also difficult as one goes to earlier developmental stages. Furthermore, no developing organism really goes through the actual endpoints of its ancestors. The problem has been discussed lucidly by G. R. de Beer (1940), with reference to morphological characters, and will not be belabored here.

Nevertheless, we should not be surprised if ontogenetic sequences in behavioral development vaguely resemble phylogenetic changes. As a method of reconstructing phylogeny, however, this correlation hardly deserves serious consideration.

POPULATION MODELS

It is also possible to make models of the evolution of behavior by artificially applying a selective force upon a captive, reproducing population.

The existence of so many behaviorally distinct varieties of domestic animals, e.g., the friendly cocker spaniel and the aggressive Doberman, attests to the effectiveness of such selection. The studies of Manning (1960) indicate how such major behavioral differences can arise from relatively minor changes. His work with two sibling species of the fruit fly, *Drosophila,* suggests they came to differ because natural selection acting upon a common ancestral population favored high activity in some environments, low activity in others. The changes in reaction threshold that then occurred served to differentiate populations in the two environments, which thereafter continued to diverge. The striking differences in mating behavior that now exist between species can thus be largely attributed to mutations that did no more than affect activity thresholds.

History in Small Populations

We noted earlier (Chap. 9) that small populations may be too homo-genous to provide the variation upon which natural selection can act. Thus, whatever (behavioral) characteristics the population possesses will remain about the same. But how do we explain the characteristics of small popula-tions such as Curio's (1961) Spanish pied flycatchers when these differ from the main population of conspecifics? One explanation involves the history not of the isolated small population but of the main one. That is, after separation, the main population continued to evolve under natural selection, while the isolated one did not. But there are two other ways in which the differences can be explained in terms of the history of the small population.

IMMIGRATION SAMPLING (FOUNDER EFFECT)

If a small sample is taken randomly from a population of anything that varies, the sample may not reflect very closely the average characteristic of the original population. This small sample may only include extreme de-viants, for instance.

There are several ways in which a small sample of animals can become isolated from the main population. A few birds can inhabit some islands (as the ancestor of Darwin's finches on the Galapagos Archipelago pre-sumably did). Or a relic population may occur when man has destroyed much of the species' original habitat in an area. This may be the origin of the montane pied flycatchers, since Spain was completely forested about 600 years ago but has now been denuded except for mountain slopes (Curio, personal communication). Or some natural event such as a retreat-ing glacier may leave behind, in "pockets," small populations of organisms that sought retreat during the glacier's advance. Thus, the founding sample of individuals may differ in behavior or morphology from the larger parent population. This origin of (possibly nonadaptive) characters is called *im-migration sampling* or the founder effect. This could also have been the origin of the seventeen behavioral peculiarities shown by the Spanish pied flycatchers (Curio, personal communication).

GENETIC DRIFT

Once isolated, a small population can change in genetic (and thus phenotypic) composition by random processes. For instance, a relatively well-adapted individual might be killed (say, by lightning) and thus leave no offspring. In larger populations such disasters occur, on the average, with equal frequency in all genotypes, and thus these random processes have little effect on the average gene composition of the population as a whole.

However, invoking this explanation of genetic drift on an *ad hoc* basis to explain characters that are seemingly without adaptive value is dangerous. It is dangerous because advantages are often subtle and thus escape even the most careful investigators. For this reason, most population geneticists have restricted the application of *genetic drift* to those characters which can be shown to be of *dis*advantage to the animals possessing them. It is only in this case that nonadaptive maintenance (see p. 211), immigration sampling, and genetic drift are convincing explanations of the occurrence of behavioral and morphological characters.

Genetic drift might also be a factor in the origin of differences in pied flycatcher populations. The effect remains to be demonstrated for a behavior pattern.

Cultural History

The foregoing discussions have dealt with historical changes in behavior that are due to alternations in gene frequencies of a population. But, when a general ability is present, especially the ability for social learning, the possibility of cultural transmission of behavior also obtains. Although firm social traditions may occur only in higher animals, such traditions may be widespread. Their historical origin then becomes a question of importance.

For instance, one may wonder why the monkeys of some islands in the Japanese archipelago wash potatoes before eating, but individuals of the same species on other islands do not. Fortuitously, the answer to this particular question is known because some Japanese anthropologists (e.g., Miyadi, 1959) observed the origin and spread of this tradition. On an island where fishermen were in the habit of giving potatoes to monkeys, one animal was seen to roll its potato in the water one day. "Friends" of the individual copied this act, and the habit spread inland along the rivers and finally throughout the entire island. Studies such as this certainly deserve the attention of future ethologists, and it may be expected that studies of tradition learning will come to represent an increasingly important area of ethology.

A Caveat: Does Behavior Evolve?

In the foregoing pages we have assumed that behavior patterns do evolve and that their history can, at least sometimes, become known. Ethologists have also argued that because some behavior patterns tend to be relatively conservative, that is, resistant to change, they are particularly good clues to the evolutionary history of their possessors. Phylogenies of spiders, pigeons, and waterfowl have been based upon this approach (Petrunke-witsch, 1926; Whitman, 1919; and Heinroth, 1910). There is a degree of circularity in this logic.

Similarities between organisms can be due to convergence, as well as to kinship. Many insects and amphibians of the tropical rain forest are green. Their hues arise from very different sources, physical structure on the one hand (hairs, refracting scales) or pigments on the other. Even pigments with similar reflectances may be composed of different molecules. Convergences may occur at a variety of levels, however, and these differences in levels are not always obvious. Thus, the colors of cryptic insects represent a functional convergence at the interspecific level, the common function being protection from predation. The functional similarity in the wings of birds, bats, and insects, on the other hand, represents a different order of adaptation, one dependent upon a larger range of conditions. The functional analogies between the excretory cells of earthworms and the nephrons of vertebrates are, in contrast, due to a narrower range of conditions. The recognition of similarities in design may thus depend heavily on the background of the investigator and the "graininess" of his perceptual film. What this implies is that the important distinction between analogies and homologies is largely dependent on prior knowledge of the phyletic relationship and physiology of the animal in question (Klopfer, 1973; also see G. Bateson, in press).

In Chap. 9 we considered the bases for behavioral convergence, the source of analogic resemblances. This chapter listed the bases for behavioral phylogeny, the explanation of homologous similarities. Do we now conclude that the two are indistinguishable? Is the issue of the distinction even important?

The issue is, we believe, an important one, both for obvious scientific as well as for political reasons. The latter have to do with the temptation to ever-widen the scope of our phylogenetic reconstructions. From the species of a genus of waterfowl ethologists (and others) have moved to the genera of an order of primates, and finally have extrapolated freely from almost any vertebrate to man.

Much of our (human) behavior has been explained or excused on the assumption that it represents a biological heritage that we can no more deny than our parentage. This has been particularly true of such patterns as those labeled "aggression," "territoriality" and "mother love." The evolutionary origins of these behavior patterns and their underlying mechanisms are understood on the basis of phylogenetic extrapolations from particular species. It is only occasionally admitted that the end point of a particular extrapolation will depend crucially on the species selected. Given the multiplicity of mechanisms that subserve common ends, even among related species, the choice of species becomes so arbitrary as to preclude any meaningful conclusions. English robins defend individual territories; Galapagos mockingbirds may have communal territories. How can we know which of these provides the more appropriate analogue or homologue of the behavior of men? D. Lehrman has asked how congressmen who opposed child-care centers would have responded had studies of motherless bonnet macaques been presented to them rather than of rhesus. Bonnet monkeys accept surrogates considerably more readily than do rhesus (Klopfer, 1973b, in press).

Which is the more appropriate model for human behavior?

The resolution of the dilemma lies, first, in the recognition that behavior patterns are rarely to be attributed solely either to historical events or present pressures. Both are, necessarily, an element in the causation of every pattern, although their relative contribution may differ. It has even been suggested that there is a complimentary relationship such that adaptability (capacity for response to the immediate situation) is inversely proportional to adaptation (the phylogenetic contribution), although it has only been in rather special and limited circumstances that a quantitative test has been possible (Gause, 1942).

Second, phylogenetic inferences respecting the origin of behavior demand independent evidence on the existence of a particular lineage (or vice versa). One cannot infer both a lineage and the evolutionary sequence of a behavior pattern!

Finally, the information from functional studies must be incorporated into phyletic analyses in such a way as to allow one to "generate the rules of the game." By this is meant that we must specify both the ecological factors and evolutionary accidents that have provided all the significant constraints on particular patterns of behavior. Having done this for an array of related species, we see that reliable generalizations applicable to others would then be expected to emerge. For instance, Crook (1970) and Eisenberg et al. (1972) have examined the relation between habitat and social structure in various primates. Observations of behavior have led to some specific predictions. Aboreal, diurnal, leaf-eating species tend toward small home ranges and troops with but a single adult male, irrespective of taxonomic relationships. Other features of their behavior are presumably independent of habitat or sociality and thus may be assumed to reflect a common history. The point of difference from the Empeid fly example cited above (p. 207) is that the behavioral constraints are being examined *en bloc*, and holistic effects noted. This is necessary for, as we hope to show in the next chapter, behavior is a process and need not depend on particularity at the molecular or cellular level. No two movements, however, repetitious, are likely to involve the same muscle or nerve cells. Instead, the behaving organism acts like a system that examines its own output, maintaining thereby a functional constancy unrelated to structural constancy, but very sensitive to changes in its environment (cf. M. C. Bateson, 1972). What does this imply for evolutionary and comparative studies of behavior?

It implies a high degree of malleability, such that the same structure (in different environments) can give rise to very different patterns of behavior, while similar patterns may be due to different structures. Thus, comparative studies of behavior are of themselves unlikely to elucidate the evolutionary sequence or to shed light on the significance of specific differences. Our studies must rather be of a molar nature, encompassing both the behavior and the ecology of species that interest us. The evolutionary game is played in an ecological theater (Hutchinson, 1965). If we want to know the rules of that game, we must observe it within that theater.

Selected Readings

Bermant, G. 1973. *Perspectives in Animal Behavior.* (This book includes a chapter by Klopfer with other critiques of the comparative method for studies of evolution.)

Wickler, W. von 1961. Ökologie und Stammesgeschichte von Verhaltensweisen. *Fortschritte der Zoologie,* **13**:303–365. (A recent monograph reviewing studies of the evolution of behavior. Although it does not delineate all the methods of study discussed above, it provides many examples of recent and older literature, and it reviews some general problems of evolutionary research.)

Chapter 11

HOW IS BEHAVIOR CONTROLLED?

External Stimuli and Internal Mechanisms

Introduction

The question, "How is a behavior pattern controlled?" might bring different answers from different kinds of biologists. An ecologist, for instance, might list environmental situations, the occurrences of which are correlated with particular kinds of behavior. A physiologist might answer in terms of the functional capacities of sense organs, brain centers, and such structures.

Since in this book we have looked upon behavior as a series of interactions between an organism and its environment, we must include both the ecologist's and physiologist's replies. The ecologist's explanation is a molar approach to behavioral control: how is a stimulus transformed (in the descriptive sense) into a response? The physiologist's approach is molecular: what are the series of transformations (and the structures that mediate them) that take place between stimulus and response? Thus, the two approaches of long-separated disciplines are complementary. Indeed, some have conceived of the nervous system as a representation of the organism's environment, thus emphasizing the value of a unitary approach to behavioral control (Young, 1964).

One presumably could list the results of these two approaches, but the list would be mere history, and it is the object of these final chapters to point the way to the future of ethology. Therefore, in the discussion to follow we shall focus on some recent studies which attempt to bring together these two approaches to the study of behavioral control. (The RECENT DEVELOPMENTS sections of nearly all chapters in Part III, particularly Chaps. 4, 5, and 6, also discuss studies on the control of behavior.)

We use the rubric *behavioral control* to include all the proximate causal factors that influence behavior, factors which often have been studied more or less in isolation. Excluded from consideration are the more or less permanent changes in behavior which take place during ontogenetic development (see Chap. 12), recognizing that this exclusion may prove to be somewhat arbitrary.

In other words, the study of behavioral control is the study of external stimulus situations that elicit, arouse, direct, or inhibit behavior. It is also the study of what has been termed motivation, drive, and the interrelationships between behavior patterns.

BEHAVIOR AS A PROBLEM OF CONTROL

The experiences of investigators in mathematics and electronics have been combined into a general theory of control called cybernetics. Of particular importance to the study of behavioral control are (1) transformations and feedback, (2) detection theory, and (3) information measurement. The application of these approaches to analysis of the control of behavior can be illustrated by a hypothetical example.

Suppose we have a simple motor act exhibited by an animal. We can then present various controlled stimuli to the animal and note the correlation between each stimulus and each response (in this case, the act or the absence of the act). Thus, the animal transforms the stimulus-input into a response-output. If the act, in turn, leads to an alteration of the environmental stimulus, then we have a second transformation; in this latter case, the act is the input, the new stimulus the output. When two such components of a system are interconnected so that the input of one is the output of the other and *vice versa*, each component is said to provide feedback to the other. For instance, an insect (stimulus input) placed before a toad elicits striking and eating (response output); the eating (response input) feeds back to the environment by altering the stimulus (now an output) to "no insect."

There are, of course, many intermediate transformations between environmental stimulus and behavioral output, and each of these can be described in terms of transformations (from stimulus energy to nerve impulses, from nerve impulse to muscle contractions, etc.). There may be an inherent uncertainty, that is, mistakes, in these transformations. To return to the simple example, sometimes the insect may be present and the toad will not eat (a *miss*); at other times the toad may exhibit feedinglike behavior in the absence of the insect (a *false alarm*). Misses and false alarms may occur at any transformation between the environmental stimulus and the motor act, and their combined effects produce variation in the stimulus-response system.

Finally, the meaningful stimulus is only one of an ensemble of possible stimuli. It is often desirable to express quantitatively the difficulty of making the correct sensory discrimination. The more possible stimuli that exist, the more difficult the discrimination of any one of these stimuli from all others. This difficulty can be expressed quantitatively in terms of the equivalent number of yes-no questions required to uniquely specify the one correct stimulus from all possible ones. Information theory refers to these questions as *bits* (binary digits) of information. It turns out that as the number of equiprobable stimuli in the ensemble increases logarithmically, the number of bits required for specification of one from the ensemble increases arithmetically.

The transformations involved in a behavior pattern usually follow this sequence: first, some object or situation in the environment emits stimuli (light, sound, chemicals, etc.) which are transduced (or transformed from one kind of energy to another) by the animal's sense organs. The sense organ not only converts stimulus energy of the environment to the chemical energy of nerve impulses, but it also filters, transducing or enhancing some stimuli and not others.

Second, further transformations occur as the nerve impulses of the receptors (outputs of the first transformation) interact through summation and inhibition processes in exceedingly complex ways to produce further nerve impulses in second-order neurons. There may then be many such transformations in various neural paths as the sensory information is combined and recombined in various ways. Internal sources of nerve activity may interact with the sensory projection systems at any level (by changing the thresholds of the receptors, by altering transformations in the central nervous system, etc.).

Third, the nerve messages resulting from these transformations form the input to muscles (and glands) which transduce the chemical nerve energy to mechanical energy or chemical energy.

Fourth, this last transformation, which results in the motor act, feeds back in some way to terminate the messages going to the effector organs. Feedback may be entirely internal (e.g., proprioceptive) and alter transformations at the second step in the sequence. Alternatively, feedback may alter the stimulus input (step one) either by changing the environment or by altering the relation of the animal to the environment.

Description of this general sequence is not, of course, new. However, the formulation of the logical steps of the series in terms of precise description and measurement and the synthesis of ecological and physiological sequences into one control system represent the new approach to behavioral control.

SOME BEHAVIORAL CONTROL SYSTEMS

Four recent studies demonstrate the approach combining external and internal steps of behavioral control.

Gulls usually lay exactly three eggs (rarely two). By removing eggs as they were laid and placing additional eggs in nests, Weidmann (1956) discovered how egg laying is controlled. Under the influence of hormonal states caused by reproductive displays, egg follicles develop sequentially in the ovary. Upon the laying of the first egg, the female commences incubating. Incubating does not affect the next most advanced follicle but may cause degeneration in the third, and it always causes degeneration in the fourth follicle. Thus, the second egg is always laid and usually the third is too, but no others. Removing the first egg prevents incubating and thus eliminates the inhibitory feedback to the ovaries. By continual removal of eggs as they were laid, Weidmann "milked" up to seven eggs from a single female. By placing a wooden dummy egg in the nest before any follicles reached the critical stage in which they were no longer inhibited by incubating feedback, egg laying was completely prevented; in fact, the duped females incubated the single wooden egg for several months before finally leaving the colony at the end of the breeding season.

As an example that illustrates another aspect of the study of behavioral control, consider the questions that may be asked of a relatively simple response. A gull chick upon hatching pecks at the beak of its parent; in response to the pecking and other stimuli from the chick, the parent regurgitates a mass of semi-digested food upon which the chick feeds. Since the chick rarely pecks at anything except the parent's beak but does not always peck at the beak when it would be possible to do so, we may ask why it pecks when it does? In other words, what factors in the external stimulus situation elicit the chick's pecking and orient the response to the parental beak? What (internal) factors move the chick to peck sometimes but not at other times when the external stimulus situation remains constant?

A preliminary answer, in crude qualitative form, was given as follows (Hailman, 1964a). The chick responds optimally to certain visual configurations, movement patterns, and colors. This can be discovered by the use of simple models presented to the chick. The parent's bill provides a nearly optimal stimulus when held in a vertical position and moved horizontally; pecking of the chick elicits vertical positioning of the beak from the parent if the latter is ready to feed. Thus, some pecking from a chick induces a stimulus situation that elicits even greater pecking, which in turn elicits regurgitation.

What factors initiate the first pecking in a previously quiescent chick? Observations from a hide reveal that the probability of a response is partially dependent on the time that has elapsed since the chick last fed, that is, on the chick's state of food deprivation. At any given time, therefore, at least two factors must be known in order to predict the probability of a response: the external stimulus situation and the internal state of readiness.

As the chick feeds, however, the probability of pecking decreases, and this change, in turn, induces a change in the parent's behavior. The parent picks up the food mass in its beak, providing a stimulus situation that again raises the pecking rate of the chick because of the chick's predilection to peck at the beak. Again, the chick eventually stops pecking, which elicits reswallowing of the food by the parent.

In sum, the parent's and chick's behavior are bound in a more or less determinate manner (Tables 1–4 in Hailman, 1964a). These observed relationships can be summarized in a diagram (Fig. 11–1), which resembles a simple version of the control diagrams that servosystem engineers use to summarize electronic circuits when the properties of the individual components are known. Here the major components are the parent and the

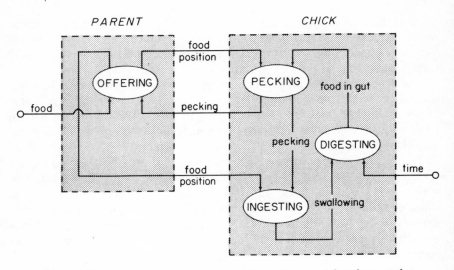

Fig. 11-1. This diagram summarizes the environmental and internal controls of pecking.

chick; observable variables which influence the behavior of major components (inputs) are represented by incoming arrows and responses (outputs) by outgoing arrows.

In gross terms, then, we have a representation of the control of pecking behavior, an example that illustrates social interactions, sensory coding, problems of drive, and other features of behavior. Exact measurements of timing, pecking rate, and other parameters will make the system more determinate, as will the discovery of other minor components that influence the system. The diagram represents a program of research based on a summary of initial observations. It is the attempt to make the dynamics of such a system determinate that may be taken as one goal of studies on the control of behavior.

A study that deals directly with internal transformations is Mittelstaedt's (1962a, 1962b) investigation of the strike behavior of the praying mantid. Mantids catch their prey by rapidly extending their forelegs, such a strike requiring from 30 to 60 thousandths of a second. This time is not sufficient to allow for corrections in aiming should the prey have moved after the initiation of the strike. The aiming of the strike is apparently dependent on an input to the compound eyes. However, the eye is not fixed either with respect to the body axis or the prey. How, then, is the strike aimed?

First, the mantid fixates a prey organism. The primary input, then, is the deviation of the prey from the eye axis. The final output is the deviation of the foreleg strike from the body axis. A correction is necessary to account for the variable relation of eye to body axis. A secondary or correcting loop apparently exists whose input stems from proprioceptors of head position. This secondary input is proportional to the extent of the deviation between eye axis and body axis. A somewhat simplified discussion of the details of this system will be found in Roeder (1963).

A final example illustrating the possible relation between feedback control systems and those centers involved in the phenomenon of consciousness is provided by the reafference model of von Holst and Mittelstaedt (1950). The behaving organism acts as if a decision to act is registered as a photographic negative; when an appropriate motor response is made, the proprioceptive input generates a photograhic positive which effectively erases the record of the decison. If something goes awry, preventing the erasure, the command to act will continue in effect. Consciousness represents an awareness of the sustained presence of the "negative" or of an incongruity between the command record (the negative) and the resultant proprioception (the "positive"). (See p. 108 for diagram and further details.)

With these brief examples of how external and internal stimuli initiate effector acts and how the acts themselves feed back to negate the initial stimuli, let us now turn to more detailed examinations of restricted aspects of the total control system.

Sensory Coding

The first step in the behavioral sequence outlined above concerned the transduction of stimulus energy from the environment to nerve impulses and the differential filtering of stimuli. The biophysical processes of transduction are not of direct interest here, except that when finally described in the various sensory systems they may be found to impose filtering properties upon the receptor. What is of direct interest is the kind and quantity of information a given sensory system can encode.

For instance, ants emerging from their mound find their way to a food source such as sugar with incredible accuracy. Wilson (1962) has shown that the ants know the direction and distance of food from a signal given by a returning forager or scout. The signal is a chemical trail. Wilson measured the inherent inaccuracy of the laid trail (as "source entropy" in bits) and the accuracy with which emerging ants follow the trail. From these measurements he could calculate the amount of information transferred about direction and distance to food. A direct quantitative comparison with similar information transferred to emerging bees by the dances of their returning companions (see Chap. 6) is possible, even though the information containing stimuli are entirely different (i.e., chemical versus tactile stimuli).

In addition to ascertaining the information content of a signal, it is of interest to determine the nature of the code employed by a sensory system. Many sensory systems are very general, being used to convey information about a variety of stimulus situations, whereas others apparently are quite specialized, having been selected for in evolution to code only one or a small number of perceptual preferences. Examples follow.

FORM PREFERENCES

Among the various sensory systems of vertebrates, the greatest advances in understanding seem currently to lie in the nature of form-vision. Hartline (1938) described the activity of third-order neurons from the eye of the frog; activity in these units could be elicited by either the onset or cessation of a light pulse. In fact, visual-system neurons can be classified as *on* cells, which respond with a burst of nerve impulses to the onset of light, analogous *off* cells, and *on-off* cells, although this classification is a gross simplification.

It has been discovered that such neurons are arranged in definite patterns in the cat's visual system (e.g., Hubel and Wiesel, 1965). Single retinal ganglion cells, for instance, can be activated by a light spot over a relatively large, but restricted portion of the retina (the *receptive field*). One kind of neuron responds to the onset in the center of the field and the cessation of light in the periphery of the field; other units are just the reverse. At higher stations in the visual system units have more complicated patterns of response, having an elongate center rather than a circular one; again the center may respond either to the onset of light (some cells) or the cessation (other cells), with the periphery behaving in the reverse fashion. The long axis of the center may be orientated in any direction, but within a given column of cells in the visual cortex all units are orientated in the same direction.

Although the way in which the cat uses sensory codes in its visually guided behavior is unknown, it is of interest to note that the cortical cells (of cats) respond to nearly the same elongate configuration that elicits maximum pecking rates from gull chicks (Hailman, 1964a). Therefore, this sort of unit might be sought electrophysiologically in the visual system of gulls as well.

At still higher levels in the cat's visual system, seven or eight neurons away from the retinal receptors, Hubel and Wiesel found *complex* and *hypercomplex* cells of several types. One of these, for instance, responds to a bar-shaped stimulus regardless of its position within the receptive field.

A comparable level of complexity to that found in the cat's brain has also been found in the eye of the frog (Maturana et al., 1960). Ganglion cells can be placed into five categories on the basis of their response to external stimuli. One of these five responds simply to the level of illumination, but the other four respond only to certain shapes moved in certain directions, e.g., a dark, convex shape moving in the direction of its convexity. The investigators speculate that such units could be "fly-detectors" that ultimately elicit a feeding-strike on the part of the frog. They also suggest, on the basis of histologically discovered connections of neurons in the retina, how receptors may be arranged to produce the complex recognition abilities of the ganglion cells. Electrical recording in higher centers of the frog's visual system revealed cells of extreme complexity, which were active only when a new object entered the visual field or while an old, fly-like object was still within the field but no longer in motion. Somewhat similar units have been discovered by Maturana (1964; Maturana and Frenk, 1963) in the visual system of the pigeon.

COLOR PREFERENCES

When an animal responds preferentially to certain spectral combinations of light independently of the light intensities, it may be said to show a color preference. We cannot assume that animals see the same phenomenal hues as human beings, but we can measure spectral preferences and their neural coding.

The most exciting advance in this area has been Muntz's (1962a, 1962b) demonstration that single units in the frog's diencephalon respond to the same wavelengths as does the whole animal during an escape response. Previous results of Granit (1942) indicate that much of the coding takes place within the retina, and Muntz (1963a, 1963b) has suggested a tentative three-receptor scheme of interaction that might code the response.

A similar scheme for coding of the gull chick's color preference was hypothesized in quantitative form (Hailman, 1964c). (See Fig. 11–2.) The cone receptor bearing a red-colored oil droplet is activated by light at any given wavelength to give a response proportional to the logarithm of the product of the intensity of light, the transmission of the oil droplet, and the absorption coefficient of the visual pigment; this receptor response provides the excitatory function of the coding scheme. Cones bearing yellow oil droplets—thought to have the same visual pigment as the "red" cones, but a different spectral response due to the differences in transmission spectra of the oil droplets—provide an inhibitory input. Other evidence suggests that cones with nearly colorless droplets may act synergistically with the "yellow" cones in providing inhibition (Hailman, 1966). A theoretical spectral response curve for some visual unit beyond the receptors themselves may be calculated on the basis of oil-droplet transmission spectra; when plotted, this theoretical curve resembles the actual spectral preference curve of the pecking gull chick. The preference could be coded

Fig. 11–2. Mechanisms of Behavior: Gull chicks prefer to peck at *red* spots over spots of other colors. The large hemispheres are meant to be the two eyes. They contain oil droplets of two colors, either red ones or yellow ones. For the sake of simplicity, the lower eye has been shown containing only the red droplets and the upper as containing only yellow ones. There are twice as many yellow as red droplets. The squarish figures, top and bottom, represent sources of light, red, yellow, and blue. The red droplets act as filters which allow only red light to pass; these droplets are only found in cells which are excitatory. The yellow droplets allow all red as well as other colors of light to pass; they are in cells which connect to brain areas that are inhibitory. Whether the CNS is "excited" or "inhibited" then depends on the ratio of "red" to "yellow" cells that are firing. (From P. H. Klopfer, *On Behavior: Instinct Is a Cheshire Cat,* © 1973. By permission of J. B. Lippincott Company.)

in the retina as early as the second neuron (bipolar cell), a possibility open to electrophysiological test.

AUDITORY PREFERENCES

Other workers have provided a powerful analysis of how the bullfrog recognizes calls of its conspecifics. Single unit recordings from the neurons originating in the sense organs revealed *simple* units with peak sensitivities around 1500 cps and "complex" units with peaks around 400 cps (Frish-kopf and Goldstein, 1963). The bullfrog's voice has considerable energy in these two ranges, with little energy in the intervening frequencies. Sound in those middle frequencies was found to inhibit activity of the complex units. Thus, for both kinds of auditory nerve cells to pass information centrally at the same time, a sound must resemble the frog's call. Perhaps there exists a higher center in auditory portions of the frog's brain which is active only when both simple and complex units are firing simultaneously.

Capranica (1965) found that bullfrogs will call in response to hearing another bullfrog. The minimum sound required to induce calling is a combination of two tones, one in the frequency range of complex units and one in the range of simple units. Energy added to the intermediate frequencies abolishes calling.

The inhibition in the case of the complex units cannot be neural inhibition because these are first order neurons, that is, they come directly from the sense organs. A team of investigators (Geisler et al., 1964) found anatomically that simple and complex units originate at different sense organs in the frog's inner ear (the basilar and amphibian papillae, respectively). The sense organ of the complex units is anatomically more complex than that of the simple units, suggesting a mechanical inhibition within the receptor itself.

A simpler case of auditory coding exists in the ears of noctuid moths. A three-celled auditory organ contains two cells maximally active in the ultrasonic frequencies produced by predatory bats. These sounds induce a variety of escape maneuvers on the part of the moth. In a tour de force of electrophysiology Roeder and Treat (1959) recorded the activity of the auditory cells in the field while the moth's ear was stimulated by the actual sounds of passing bats.

TRAINED DISCRIMINATION

All of the perceptual preferences discussed to this point appear to be characteristic of the species; in the examples of gull chick's pecking and frog escape behavior, the color preferences have been demonstrated in young animals that were more or less deprived of all sensory experience. Clearly, preferences for certain stimulus configurations can also be learned later in life. Although there exist in the psychological literature many studies correlating general activity of gross portions of the brain with various training stimuli, relatively few single-unit studies which would reveal coding mechanisms have been made (but see Pribram, 1971).

An illuminating example is the work of Erickson (e.g., 1963) on taste preferences in rats. Rats can be trained to accept or avoid water with certain chemicals added so that it is possible to find which chemicals are discriminated and which are not. The discrimination training can be accomplished precisely enough to scale the relative similarities in taste among a large array of salts, acids, and other chemicals. In this situation, asking how a trained preference is coded is bound up with the question of how the discrimination between the training chemical and all others is coded.

Erickson recorded activity of third-order neurons in the taste system and found that nearly every unit responds to nearly every chemical. Suppose one arranges a number of units on the abscissa of a graph. (Actually, there is no basis for assuming that the units form a unidimensional continuum, or any continuum, for that matter; however, the coding scheme is easier to demonstrate as a series of graphic curves than as a series of bar graphs.) Then graph on the ordinate the response rate of each fiber to a particular chemical and connect the graphed points to form a *cross-unit* curve for that chemical. When this procedure is repeated for a series of chemicals, it is found that some chemicals have rather similar curves (Fig.

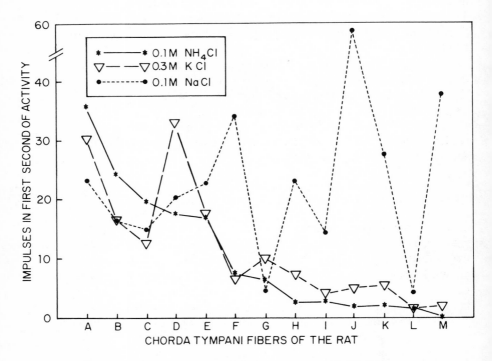

Fig. 11–3. Firing patterns of various neurons (*A-M*). The neurons are arranged in order of decreasing sensitivity to NH$_4$Cl. Note that the pattern of response to KCl is similar, but that to NaCl is different. (After Erickson, 1963.)

11–3). Erickson found that chemicals with similar curves are discriminated poorly by rats, whereas chemicals with quite dissimilar curves are easily discriminated. It is not known how these cross-unit firing patterns for chemicals are coded by the receptors, nor is it known how the information is further processed centrally. The study does, though, open the door to investigation of the coding of trained perceptual discriminations and preferences.

If a pigeon is rewarded for pecking at a particular wavelength of light, and then its wavelength preference recorded subsequent to training, a spectral generalization curve or preference is obtained (Guttman and Kalish, 1956). Such curves may be obtained anywhere in the visible spectrum, and all such curves are similar under similar training conditions: the peak of the pecking rate is located at the training wavelength, the peck rates decrease linearly and symmetrically on either side of the training wavelength as a function of spectral distance from it.

Granit (e.g., 1947, 1955) and his followers have recorded the spectral response of ganglion (third-order) cells of the retina and found in many animals curves of broad spectral sensitivity from some cells (dominators) and curves of narrow spectral sensitivity in others (modulators). Although an animal has but one or two dominators, it has many modulators located throughout the spectrum. Granit's interpretations of his findings have rested upon the way in which modulators cluster into a few spectral groups. An interesting property of these units has been overlooked by such emphasis. Each curve has a peak response at a wavelength and declines in linear, symmetrical fashion on either side of that wavelength (e.g., Donner, 1953, for the pigeon).

Our interpretation of these findings is that color training in the pigeon somehow selects the information of the modulator whose spectral peak is closest to the training wavelength. The training conditions (e.g., reinforcement schedule) may alter the exact slope of the generalization curve found behaviorally so that it has a different half-band width from the selected modulator curve, for instance, a process which may take place in more central parts of the pigeon's brain. However, the basic similarity between the two functions is always manifest: approximately linear, symmetrical curves, identical throughout the spectrum. The fact that modulators are more closely spaced in some parts of the spectrum than in others may be related to discrimination functions (curves of "just noticeable differences"), but this suggestion has not been worked out quantitatively.

A CAVEAT

It would be a mistake to conclude this section with the implication that knowledge of sensory and perceptual capacities is sufficient for an understanding of an animal's behavior. There is, in fact, much more to the study. More of the details will appear in the pages that follow, but an anticipatory note is desirable: there are constraints on perception and learning which effectively select or modify sensory inputs in ways not predictable on the basis of electrophysiological (and similar) data. Rats may learn a maze-alternation habit when the "reward" is food but not when it is water. Suppose only water had been used in a training procedure; what would one have likely concluded? Rats may also learn to avoid a particular food when the food induces nausea. The avoidance does not develop if the food is associated with an electric shock. In an imprinting situation, newly hatched ducklings may treat members of a pair of models as indistinguishable, even though these same ducklings will distinguish them in a maze (Klopfer, 1973). "The Skinnerian assumption that all discriminable stimuli can become discriminative stimuli . . . [is unfounded]" (Boakes and Halliday, 1970).

CONCLUSIONS

When physiological studies of sensory systems are intercalated with studies of stimuli that elicit responses of the whole animal, the result is a fruitful approach to understanding the afferent part of a behavioral control system. The *releasing mechanism* of classical ethology (cf. Chap. 2) then corresponds to the entire sensory system up to and including the level at which units occur whose response to stimuli corresponds in a one-to-one fashion with the behavior of the whole animal. Muntz' diencephalic units in the frog illustrate this well.

The Total Stimulus Situation

The foregoing section dealt with the stimulus input as if the stimulus object were the only relevant component. A few examples will now show how misleading such an assumption may be and further justify the final caveat.

BEHAVIORAL CONTEXT

It has been found (Grunt and Young, 1952; Fisher, 1958) that copulation responses by male mammals of several species tend to wane when the animal is confined with the same female. However, if a new female is introduced, copulation returns to its initial high level. Hale and Almquist

(1960) showed that this return is not necessarily dependent on the change in stimulus object. Bulls show the same phenomenon when servicing cows tethered to a post in a field; a new cow tied to the same post as the first one elicits a resurgence in copulatory activity. However, the resurgence can also be brought about merely by moving the first cow a few feet to a different post. In other words, altering the stimulus background in this situation is as effective as altering the stimulus object.

A different sort of example is provided by Smith (1963). He found that the kingbird flycatcher gives a "kitter" call in two quite different contexts (approaching a perch and when in consort with the female). The other side of the coin is that the kitter, a vocal display, means different things to another bird, depending on the behavioral context in which the displaying bird calls. The role of context in communication at all levels has become increasingly emphasized (note Sebeok, 1968).

CONTEXT CODING

If the same stimulus elicits different responses depending on the context, the responses may be elicited by different total stimulus situations and the coding mechanisms for the differences in background should be as amenable to physiological coding studies as are the stimulus objects. Unfortunately, little attention has been paid to such phenomena. One simple example may be cited.

Although a gull chick pecks preferentially at certain colors (see previous sections), the rate of pecking is not uniquely determined by stimulus object alone. Hailman (1966) found that the color of the background field was as important in determining the response as the color of the stimulus object. It turns out that the color preference for the background is the spectral mirror image of that for the stimulus object. The coding of this situation may be involved with the receptive fields which presumably code the stimulus object since there is electrophysiological evidence from other vertebrates that receptive field centers have different responses to colors from those of receptive field peripheries (see literature cited in Hailman, 1966).

ATTENTIVENESS

When different, particularly conflicting, sensory information is being received simultaneously, an animal may restrict its attention to one set of stimuli. Horn (1965) has recently reviewed some literature on such phenomena and found two kinds of mechanisms that contribute to attentiveness. One is involved with gross morphological changes in receptor organs which restrict the inflow of information, for instance, changes in directionality of external pinnae which gather sound or changes in accommodation of the eye. These changes tend to focus, spatially, upon some source of stimulation. The other mechanism is a sort of switching circuit (e.g., in the tectotegmental region of cats) which allows only certain sensory inputs to have access to higher neural centers. The possibility of central feedback that alters receptor neural firing patterns is considered by Horn to be unproven since relevant variables have been uncontrolled in studies purporting to demonstrate such phenomena. Yet, there do exist centrifugal fibers to sense organs (e.g., to the pigeon's retina) so that this possibility cannot be ruled out.

Attentiveness as measured operationally is rather closely allied to the subjective phenomenon we know as "expectancy." Pribram and Melges (1969) have proposed a model that illustrates the manner in which the past experience of an organism (which may alter expectancy) may feed back upon the sensory input.

This model assumes that arousal, expectancy, and emotional states in general result from a discrepancy between some established pattern of nervous activity and an external change that alters that pattern. The emotions themselves are categorized as belonging to two sets. One set is a reflection of the state of disorganization of ordinarily stable configurations ("concurrent emotions"). The other is a reflection of the processes by means of which the disorganization or imbalance is redressed through a regulation of inputs ("prospective emotions") (see Fig. IV–8, Pribram and Melges, 1969).

"Input" represents any kind of stimulus. Over a period of time it becomes associated with a particular configuration of neural activity. This provides a baseline. Whenever the input that initially established the baseline recurs, a comparison is made. Does the input produce the same configurations as it did previously? If so, there is congruity, and matters end there. The process involved is the test for congruity, representing the concurrent emotion.

If the input alters from the expected, arousal results. There follows an appraisal of the significance of the incongruity. Perhaps the incongruous input is nonetheless seen as desirable, in which case the test template will be altered, or the baseline changed. If the input is judged undesirable, it may be reduced or eliminated. This process of input gating gives rise to the prospective emotions.

Gating reduces the information reaching the test center. It therefore enhances redundancy. The speed with which the internal processing occurs must, therefore, speed up. That means internal activities become fast relative to external activities. If there are two clocks, an internal clock and an external clock, the latter keeping some constant or absolute rate of time, and the internal one is speeded up, then with respect to the external one, time is seen to go more slowly; time drags.

Incorporation, on the other hand, implies an increase in the amount of information being processed and a reduction in redundancy which must result in the neural flow. With internal activity flowing fast relative to external time, time must race on, and, again, is this not a characteristic of those situations in which incongruities are appraised as highly relevant and desirable?

How does this relate to learning? The behavioral changes taken as evidence for the occurrence of learning imply a change in the valence of some particular input, or, to put it differently, in the character of a percept induced by a stimulus: there is thus no way to disassociate the phenomenon we term "learning" from perception. The Pribram-Melges TOTE model relates perception to emotion, but the intimacy of perception and learning allows it to apply equally well to learning. More on this below.

CONCLUSIONS

These brief remarks concerning the importance of the total stimulus situation not only demonstrate the difficulty of defining the real sensory input of a control system that underlies behavior, but they also lead quickly into questions of central integration of sensory information. In the following three major sections we shall present some views concerning the organization of behavior and the integrative mechanisms underlying such organization.

Drives and Behavioral Transductions

In a hypothetical, very simple organism the coded information that is sent to sensory areas in the brain might merely activate neural pathways that code a given motor response. For each different response there would be a different motor code activated by a separate sensory input. In reality, few of the complex adaptive behavior patterns manifested by animals in nature are so overtly determinate; animals are rarely stimulus-response automatons. It is because we are unable to predict the occurrence of a behavior pattern merely by complete knowledge of the immediate environmental stimulus situation that the problems of "drive" and "motivation" arise.

A quantitative illustration of this point: black-headed gulls which have just laid the first egg spend a mean time of 24.2 minutes per half-hour watching incubating, and they complete (with a quivering action) 23.7% of their resettling movements on the nest. Later when they have a full clutch of three eggs, the statistics are much higher: 29.6 minutes and 82.6%. Beer (1961, 1962a and b) took similar data from birds in mid-incubation that had lost eggs through predation, so had only a clutch of one egg, and found higher figures (29.6 minutes and 62.1%) than those from one-egg birds during laying, but lower than that from the full-clutch birds. Furthermore, the reverse experiment, adding two eggs just after the laying of the first, raised the measures somewhat (28.2 minutes and 46.3%) over the normal one-egg situation, but not enough to equal the normal incubation of three-egg clutches during ordinary incubation. Clearly, incubation behavior in the gull is not controlled uniquely by the external stimulus situation (i.e., number of eggs) but depends also on the time during the breeding cycle, that is, depends presumably on factors of internal readiness within the bird itself. It is these unknown internal factors that are labeled "drive."

Drive is one of the knottiest problems of ethology, for the term embraces so many different concepts. One of the clearest critics of the concepts of drives has been Hinde (1970) whose works should be consulted for details not documented here (see Chap. 2).

Drives are postulated internal states of activity which interact with the exterosensory input to bring about behavioral responses. It has been presumed that these drives are the result of long-term, nonspecific stimuli acting either directly on the nervous system or by releasing hormones, or both. Some authors have also postulated spontaneous states of activation which somehow arise independently of environmental influences. Closely related to this general view of drive are vacuum activities (overt spontaneous behavior), displacement activities, and other problems deriving from the failure of the organism to act in a one-to-one relationship with immediate stimuli (see discussion later in this chapter).

Many, if not all, of these problems reflect quite complex interactions of the central and peripheral nervous systems. Some, if not all, of these problems may eventually yield to a complete behavioral description—yield, that is, when the level of behavioral observation and measurement is fine-grain enough, when the total immediate stimulus situation can be accurately assessed, when the longer-term stimulus situations can be measured, and when the time scale of measurement is large but the temporal unit of measurement is small. Given such a detailed description perhaps we can predict accurately the behavior from a knowledge of the environment and the preceding behavior of the animal. Such a description would then yield models of the operations of the nervous system. We indicate below only a few examples of the attempts being made at such descriptions of behavior (see also Fentress, 1973; Simpson, 1973; Slater, 1973).

THE "FEW DRIVES" THEORY

The naturalistic study of behavior (see Chap. 2) led to a classification of behavior patterns based largely upon function: fighting, sex, and so on. This functional classification, in turn, became the basis for postulating a few unitary drives activating the responses in functional categories. In fact, it has been more or less assumed, even recently (e.g., Moynihan, 1962; Kruijt, 1964), that most or all displays are underlain by combinations of three drives: attack, fleeing, and sex.

Tinbergen (1959) articulated three criteria for assessing the relative "amounts" of such drives activating a particular display: (1) the similarities in physical form between the display and unmitigated attack (or fleeing or copulation); (2) similarity in the behavioral context in which the display and the "pure" response occur; and (3) associations between the display and the response in time. Brown (1964) enlarged the criteria to include higher orders of all categories; for instance, if a display were followed by attack, and another display preceded the first, then the other display could also be said to be activated by the attack drive, even though it was never followed by attack.

We wish to make two important points about these criteria and any similar schemes. First, if they are useful, it is more in assessing phylogenetic relationships than in assessing motivational ones (but recall the caveat of p. 229). Similarity in form is the principal criterion that the comparative morphologist (or ethologist) uses to infer common origins. However, the fact that two motor responses look similar only very indirectly suggests that their neurophysiological basis is similar. After all, an animal may approach a potential mate, an enemy, or food, but few ethologists would suggest that these similar approach patterns were activated by the same drive.

Similarly, the behavioral context is not a particularly good criterion for motivational similarities. If the external stimuli were exactly the same, for instance, the differences in observed behavior would be attributed to *different* activating drives (or at best a random process), exactly the opposite conclusion supposedly derived from the criterion! But, if the stimulus situations are different, why cannot the differences in the responses be attributed directly to the differences in environmental stimuli with no need for an intervening drive?

Tinbergen's last criterion seems the most relevant. If two behavior patterns tend to be given to the same stimulus situation, or tend to accompany one another in a variety of stimulus situations, their association could be due to common sources of central activity. This criterion has some of the pitfalls of the previous one, but on the whole it is more intuitively satisfying because of the assumption that the nature of drive is not only general (i.e., it activates many different responses), but it is also relatively long-lasting. For instance, if the stimulus situation changes, and the change is not immediately followed by a change in response type, one assumes that the internal motivational state is still dictating behavior. At a later time, when the response does change, the change is not due (in the immediate sense) to the previously changed stimulus situation, except indirectly through a change in drive states. Suppose there are differences between the latencies of transitions between responses; those that change most rapidly might be assumed to share more common pathways of activity so that less of an internal change is necessary to effect the transition.

This line of argument leads to the second point concerning Tinbergen's criteria. If the temporal association of responses best reflects internal states, why must this operational criterion lead to the conclusion of a few drives? In other words, it is not a complete description merely to measure the association of display A with attack, escape, and copulation. The association of display A with all other behavioral responses could also be measured. Would the result reveal a few groups of associations corresponding to functional categories of behavior? One possibility would be that there exist major groupings of related displays (which might, or might not, be assigned to attack, escape, and so forth) and within these major groupings would occur minor groupings of more closely related displays. Ultimately, a hierarchial system reminiscent of early ethological postulates (see Chap. 2) might result.

The reader will probably now have reached the obvious conclusion. If we can effect such a total description for the prediction of behavior, of what use are the unobservable intervening concepts of drives? Can it be that with this question ethology is on a threshold of an operationalism analogous to that confronting Watson at the founding of behaviorism (cf. Chap. 3)?

Perhaps the present dilemma, if there is one, has arisen from confusion in attempting to define with a single conceptual framework two disparate causes and origins of behavior: function (cf. Chap. 9) and control. Behavior patterns ultimately serving the same function will often be temporally associated, at least over a gross time scale. In a gross way, then, they will be motivationally related. But the extension of these general functional relationships to hypotheses about behavioral organization and neurophysiology is misleading. There do not necessarily exist brain centers the stimulation of which yields "sexual behavior" or any other very general category. Our prediction is that the functional categories will remain; they are useful in their place. However, the study of motivation will shift to a new, fine-grain analysis of behavioral responses related by operational criteria. Since temporal association is the easiest of such criteria to measure accurately, it comes as no surprise that recent studies have concentrated on this area (see also Simpson, 1973).

TRANSITIONAL PROBABILITIES

Responses may be related in time in two principal ways: sequence and temporal patterning. The first is easier to measure and therefore was studied earlier. Andrew (1956) charted the frequency with which given preeninglike responses were followed by the same response and by each of the other preening responses. Preening, stretching, or bill wiping tended to be repeated at a frequency greater than that expected by chance alone. By eliminating such cases, that is, by restricting analysis only to sequences in which a change in response type occurred, Andrew found two main groups of activities: scratching, preening, and feather setting; stretching, rest, and locomotion. Bill wiping occurred in isolation from both groups.

However, the situation proved to be somewhat more complex than could be handled by a *few drives* concept, even when operationally defined in this way. For instance, stretching often led to movement, scratching to yawning, and resting to stretching, whereas the reverse sequences were no more frequent than expected by chance. This phenomenon of unidirectional behavioral chains would seem to preclude a simple hypothesis of motivational centers merely activating several responses simultaneously.

Hailman (unpublished) repeated Andrew's study in a number of other emberizine species preening in a slightly different behavioral context and could find few transitional probabilities reliably greater than expected by chance in a large series of observations (except that a response tends to precede and follow the same response). In other words, in one situation preeninglike responses change to other preeninglike responses randomly. This result may indicate that differences in context, behavioral situation, or external stimuli modulate transitional probabilities. Since the presumed external tactile, and possibly chemical or thermal, stimuli that elicit the preening may be altered by it, these studies do not firmly control the role of the stimuli during the preening bouts.

This kind of study of transitional frequencies and transitional probabilities is now becoming quite popular (e.g., Wiepkema, 1961). Suppose, however, that not just the immediately preceding behavior but rather a certain sequence of preceding responses determines (statistically) the response to follow. Nelson (1964) and Altmann (1965) pursued this possibility. The male fish *Corynopoma* goes through a series of courtship actions before the female, which does little displaying herself. In this relatively constant situation, the male alternates a repertoire of responses, the sequence of which was noted by Nelson. Analysis showed that, within an operationally defined sequence of courtship actions, any given response was nearly completely determined by the preceding response; that is, the behavior resembles a first-order Markov chain and could not be described with increased predictability by knowledge of the sequence of preceding events (and see Slater, 1973). An opposing view of the generality of this result is noted below (Holistic Analysis).

GENERAL TEMPORAL ASSOCIATION

The most general approach to temporal associations between behavior patterns is to measure the variation in responsiveness of a number of responses as a function of time. For instance, Beer (1962a) measured the probability of an egg-rolling response in gulls throughout the breeding season, utilizing a standard set of model eggs. He also measured the tendency of birds to incubate egg models (Beer, 1961). Over a gross time span, these responses were co-variant with time. However, over short time spans, co-variance was not apparent.

This example indicates that a broad temporal view must be combined with a fine temporal analysis to describe fully temporal associations. It remains to be seen if behavior patterns co-variant on a small time scale remain co-variant on a large one as well; it is not logically necessary that they do so. But if they do so, a temporal hierarchy of behavior patterns might be found to exist. That is, many groups of responses that co-vary over a short time span might be found; over a longer period some of these groups might tend to be correlated, forming higher-level groups; over still longer periods, the higher-level groups might themselves be grouped; and so on. Such a scheme might finally yield something comparable to the intuitive few drives and hierarchy concepts of classical ethology (see Chap. 2).

TEMPORAL PATTERNING

The two foregoing methods of assessing relationships between responses suggest a combination of approaches. It should be possible to assess the *latency* of transition between two responses, that is, their temporal patterning. Nelson (1964) cites the surprisingly few earlier attempts at this sort of analysis and then provides some interesting examples.

Take for example the transition between two courtship actions from his male fish mentioned earlier. Assume that chasing and extending are parts of a first-order Markov process as noted above, such that one tends to follow the other (if it does not follow itself). We may then ask, "When in time after one response does the next one tend to occur?" Four interresponse intervals can be analyzed: chase-chase, chase-extend, extend-chase, and extend-extend. To analyze these intervals, Nelson used the familiar "survivorship curves" of demography, semi-logarithmic plots of cumulative frequency versus the length of the interval. Responses that occur randomly after the preceding response will have survivorship curves which graph as straight lines.

What the data showed, in this case, was that the curves were not straight lines but were concave curves with two principal segments, such that responses tended to occur frequently within 10 or 15 seconds of the last response and then occurred with random probability thereafter. However, two different behavioral contexts could be distinguished: actions while the belly spot of the male was dark and while it was light. Plotted separately, survivorship curves for these two situations yielded two families of nearly straight lines, demonstrating the compound nature of the first curves. Intervals between extend-chase and chase-extend had the same distribution within one context (in fact, all curves have essentially identical slopes). When the spot is black, response transitions occur randomly in time but on a much shorter time scale than when the belly spot is light. Therefore, although the sequence of actions during courtship in this species is largely determined by a first-order Markov process, the responses exhibit temporal independence from one another. When the interresponse interval is shortened by the behavioral context, temporal independence is maintained; the mean interval length only is affected, not the temporal relationship between responses.

This temporal independence need not occur with all behavior. In the male courtship of a related fish, *Glandulocauda*, Nelson (1965) found that gulping of air during bouts of hovering display had a sharply convex survivorship curve, indicating that the gulping behavior was quite periodic, tending to occur about every 10 seconds or less. However, when the fish was engaged in chasing, intergulp intervals were distributed randomly in time. Apparently, the air received from gulps is utilized in sound production at a steady rate during hovering but not so during chasing.

The "fine-grain" temporal analysis, of which these few results are examples, clearly promises to yield more accurate and testable hypotheses about central processes than do vague theories concerning a few drives.

HOLISTIC ANALYSIS

Promising as fine-grained temporal analyses appear to be, occasionally a question has arisen on the validity of their underlying assumptions. The question derives from evidence that there can be considerable variability in the elements that make up a pattern even while the pattern remains constant. A striking example of this is provided by the work of Golani (1973) and Golani and Mendelssohn (1971) on the precopulatory displays of jackals. The elements of these displays are fixed and the overall pattern appears stable. Actually, detailed ethograms show that the sequence, or temporal relations of the elements, does vary according to the context in which they occur. "Context" includes the age of the animals, the length of the period they know each other, time of year, and other variables. "Each new pair, each season, and even each day," writes Golani, "provides new contexts which are also changed accordingly . . . each day provides new contexts that have never occurred before. Accordingly, there is a continuous change in extrinsic significance." Golani goes on to argue that analyses that are based on motivational or functional assumptions (extrinsic significance) must be distinguished from those that deal with the relatedness of the events or elements comprising a display to each other (their intrinsic significance). Structural intricacies should not be dissolved, for they may be by extrinsic analysis (with motivational, or even stochastic concepts), but should be retained in order to allow discovery of the nature of shifts from one context to another. Clearly, Golani rejects the notion of an "anatomy" of communication, in which the job is to specify the (finite) meanings of particular displays. An entirely different methodology is thus required, which Golani has built upon a nonmetric computer technique (Multidimensional Scalogram Analysis, see Guttman et al., 1969).

One aspect of this theme has found support by K. Nelson, who was at one time an initiator of Markov (e.g., extrinsic) analysis (see above). "I do not suppose," he now (1973) writes, "that description in terms of a Markov process model will in itself lead to very great understanding of the behavioral organization; and it may in fact—I speak from personal experience—blind us to a different, possibly more complex but ultimately more realistic model."

Other models considered by Nelson are derived from the TOTE system (see p. 232), Miller et al. (1960), Harman (1964), and Reiss' (1964) models of reciprocally inhibiting neuromimes. We need not consider their intricacies here, except that they too imply a distinction between the temporal and structural organization of behavior.

TRANSFORMATIONS WITH INPUT

The foregoing examples demonstrate some of the temporal relationships between ongoing responses in one individual during relatively constant stimulation. The operational approach avoids some of the problems of drive concepts but leaves in its place only an operational description, albeit quite an accurate one. To stop at such a point in the analysis of behavioral control would be like quitting after one had learned how a television set works when it is receiving a picture of a white or black wall. We really wish to know how it receives complex pictures of black and white components. The question is, can we carry the analysis further to include a description of the system under variable input, that is, when the animal experiences various stimuli?

Returning to the courtship of *Corynopoma* (Nelson, 1964), recall that the female does little displaying. However, should she approach the male during the interval between his courtship sequences, the next sequence tends to begin with more extending and quivering and less chasing than if she did not so approach. The female may also respond by nipping at the male or following him. Her responses appear to be dependent on the cumulative effect of the male's display, particularly upon bouts of twitching. When the female does respond, the action predictably alters the probability distribution of male displays in such a way as to lead to fertilization.

Spontaneity and Displacement

Two concepts that continually arise in behavioral studies and are closely related to drive concepts and central processes of behavior are *spontaneity* (sometimes called vacuum activities) and *displacement action*. They have often been confused. The first postulates some sort of build-up of motivation or energy after performance of an act, a build-up which increasingly raises the probability of a repetition of the act when the appropriate stimulus is encountered; finally, the build-up somehow "bursts the seams" and the activity appears in absence of the proper external stimulus. This kind of behavior led to the postulation of reaction specific energy (see Chap. 2).

The related notion of displacement is based on observed behavior patterns whose performance appears to be "out of context." The implied neurophysiological model is similar; external stimuli usually leading to a specific act are opposed (by other stimuli that are directly incompatible or that elicit responses incompatible with the other response system being stimulated). The dammed up energy, which cannot escape into the usual motor pathways, diverts and flows through some other response pathway (cf. Chap. 2).

Compared with the above view of motivational problems, classical treatments of spontaneity and displacement seem quite out of tune with an operational and neurophysiological approach to behavior. Can the behavior patterns which gave rise to these problems be treated in an operational manner leading to a descriptive control system, such as in motivation?

SPONTANEITY

Schleidt (1964*a*) has accomplished a fine-grain analysis of temporal patterning of the turkey's spontaneous gobbling. Interresponse intervals of a bird in a standard, somewhat deprived environment, tend to be distributed according to an approximate Poisson process, that is, random with respect to time. This is the same sort of result Nelson found in the courtship responses of fish during constant stimulation (although he also found first-order Markov chains between different responses). Thus, a sort of continuum is established between vacuum activities and behavior usually not classified as such. The same problems about neural mechanisms apply to both sorts of behavior. Schleidt (1964) has postulated a two-component system. One component increases in activity (builds up) continuously but is continually reset to a baseline with the performance of the spontaneous activity. The other component is a random process, or better, a variable internal trigger for the responses, whose activity level varies nearly randomly.

The hypothetical system is no more satisfying than older concepts of vacuum activities, except that it is described more precisely and therefore gives a more definite mechanism to be sought by physiological studies. Since even the firing pattern of single nerves can correspond to a random or Poisson distribution, it seems quite likely that some temporally covariant mechanisms for gobbling may be discovered with diligent search. Its discovery will not, of course, clarify the ultimate mystery of random neural processes, but perhaps we can learn more about the nature of the build-up and release that must somehow underlie spontaneous activity.

DISPLACEMENT ACTIVITY

As discussed in Chap. 2, the original notion of a displacement activity was a response which occurred out of context when the animal was stimulated simultaneously to make two other, mutually incompatible responses. This functionally irrelevant activity gave rise to various motivational concepts which we shall not deal with here since it is our aim to derive a description of behavior without recourse to such vague concepts. We can, however, review briefly some explanations proposed to explain what happens behaviorally when conflicting stimuli are received by an animal, or when the animal shows an irrelevant activity.

1. When competing stimuli are received, the animal may restrict its attention, as explained in a previous section, to one set of such stimuli and respond accordingly, much as if the other set of stimuli did not exist. This is probably a very widespread phenomenon and one rarely segregated for specific study.
2. When two responses are being simultaneously activated, an intermediate form of response may be given. This situation presents no prob-

lems outside of the usual ones concerning behavioral control and probably would never even be termed a displacement activity unless the compromise response were evidently maladaptive.

3. The animal stimulated may perform both appropriate responses but in alternation with one another, either in individual performances or bouts of performances. This situation poses some interesting neurophysiological problems. For instance, does the animal's attentiveness alternate between the two stimuli? Or does the performance of one behavior pattern raise its threshold above that of the other so that the second follows the first until its threshold again exceeds the continually decreasing threshold of the first? For instance, Tugendhat (1960) provided both food and an electric shock to fish entering a certain portion of an aquarium so that stimuli for both approach and feeding on one hand and fleeing on the other hand occurred simultaneously to the fish, once it had learned of the possibility of shock. Although the situation is really more complex than this, approach and fleeing did tend to alternate, and no irrelevant behavior was observed.

   Beer (1962a) found a similar situation. A gull with a full clutch of eggs was provided with an egg model secured adjacent to the nest. The bird alternately incubated and attempted to roll the pegged model into the nest, but no irrelevant behavior was observed.

4. One of the original examples of a displacement activity (cf. Chap. 2) was "displacement nest building" of the herring gull. Engaged in a territorial dispute against a formidable rival, the gull might turn to peck at, grasp, and sometimes toss over its shoulder bits of grass. Since responses like this are seen during nesting behavior, the ostensible explanation for such behavior was that drives to attack a rival and also to flee from a stronger opponent were simultaneously activated. This activation somehow was rechanneled into nest building.

   Bastock et al. (1953) offered a more parsimonious explanation. The gull merely redirects its activated attack to an inanimate object. That the displacement activity is really attack was indicated by the similarity in physical form with the usual attack at an opponent; gulls peck down at and pull on the plumage of other gulls during fighting. Any time a strange object is in the bill, it is rejected by a simple head toss. No extraordinary relationships with nest building need be postulated. This example emphasizes the need for an accurate description of all activities so that responses may be recognized in a variety of situations. Most animals have only a limited number of available motor patterns; some will perform different functions in different situations and can be activated by quite different stimuli.

   Redirection may be regarded as an example of total stimulus situations, in which the whole behavioral context supplies the stimulus configuration (see previous sections). A weak opponent elicits attack directed to the opponent; a strong opponent elicits a venting of attack on other objects. It is still legitimate to ask why attack must be vented i.e., of what selective advantage is this solution as opposed to other theoretically possible solutions (such as psychological sublimation)? A partial answer to this question about function may in fact come ultimately from neurophysiological studies. If the activation of neural circuitry for attack cannot be ignored through changes in attentiveness or other processes, perhaps it is disruptive to the entire nervous system if not vented in some response. Or, perhaps the "vented" attack also acts as a social signal; that is, it is a display in itself which indicates to the opponent (in the case of the herring gull) a high probability of attack by the displayer if the situation should alter slightly in his favor.

5. Yet another explanation of displacementlike behavior was articulated by Lind (1959). He noted that the transition from one incipient, "in context" behavior pattern to the displacement pattern often involved a motor movement common to both behavior patterns. From these observations he postulated that the proprioceptive information from the first motor pattern facilitated the performance of sequential elements from the second pattern since the two patterns share the *transitional action.*

6. Another physiologically oriented theory of displacement comes from the studies of Morris (1957) and Andrew (1956). Here the idea is that hostile and sexual encounters evoke certain autonomic responses, as well as the motor movements of displays observer. Some of these autonomic responses, e.g., piloerection, will produce stimuli that elicit other behavior, such as preening. Therefore, preening appears to occur as an irrelevant behavior pattern in fighting and courtship.

7. Finally, two studies of "displacement preening" conclude that it, too, can be explained as a simpler kind of interaction than originally conceived (van Iersel and Bol, 1958; Rowell, 1961). When two behavior patterns, impossible to execute simultaneously, are nevertheless simultaneously evoked, preening may result. The students of displacement preening established that external stimuli for ordinary preening (ruffled feathers, dust in the plumage, etc.) are always accumulating. Preening usually occurs when the animal is doing nothing else of importance, that is, during rest periods. Preening is thus actively inhibited during activities that must be performed immediately (fighting, courting, etc.). A possible explanation of displacement preening is then that the simultaneous stimulation of two of these other activities somehow removes the inhibition on preening since the intensity

of displacement preening is a function of the external stimuli that elicit normal preening.

All these explanations of apparent displacement activity do not combine to deny the existence of displacement *sensu stricto*. In fact, in a following section on electrical stimulation of the brain, some evidence for "true displacement" will be discussed. The present studies do show, however, that many ostensible cases of displacement can be more parsimoniously interpreted, and in terms more in keeping with present physiological knowledge. More significantly, the speculations of Golani (1973) and Nelson (1973) remind us that we may even be deceiving ourselves in the isolation from the whole of those acts we call "displaced."

Control Mechanisms of Drive States

The previous sections on motivational states described some operational methods now being used to describe and elucidate the dynamics of central interactions. In this section we take several examples of studies on actual neuroendrocrine mechanisms involved in drive states.

HUNGER AND FEEDING

Eating is a classical example of a behavior pattern under control through feedback. An animal deprived of food not only eats more readily and ingests more food than a similar animal not so deprived but also will work in order to obtain food (i.e., food is a reinforcement for learning in a hungry animal). Teitelbaum, his co-workers, and others have taken considerable strides in identifying the actual neural mechanisms of the hypothalamus that are responsible for the hunger drive in rats (e.g., Teitelbaum, 1964; Teitelbaum and Epstein, 1962, 1963; Hoebel and Teitelbaum, 1962).

If the ventromedial portion of the hypothalamus is destroyed by experimental lesions, rats become obese and are observed to overeat. However, these hyperphagic rats will not work hard in order to obtain food (as measured by their bar-pressing rates in a Skinner box); nor will they continue to eat if bitter substances (e.g., quinine) are added to their food. In other words, such a rat overeats, not because it is hungry in the usual sense of the word, but rather because it fails to become satiated by the intake of food. The controlling feedback somehow fails.

Teitelbaum found that such hyperphagic animals just keep eating longer than normal rats, but they do not take more frequent meals. Any normal rat made artificially obese (by insulin injections that cause overeating) will eat very little when the treatment is removed until it returns to its normal weight. Obesity also acts to restrict feeding somewhat in the rats with ventromedial hypothalamic lesions but not until they are very obese, at which point they no longer overeat, but do remain obese. As expected, an animal made obese prior to the lesions does not continue to overeat after the lesions, nor does it undereat in order to regain its original weight; it just remains obese.

These results suggest that the ventromedial hypothalamus is a neural mechanism that inhibits feeding; when it is destroyed, the animal becomes very obese before restricting food intake. In other words, in addition to the well-known short-term satiety that restricts intake at a single meal, there exists a long-term inhibition of feeding due to obesity's working through the ventromedial hypothalamus.

If, instead of the medial portion, the lateral hypothalamus is destroyed, rats stop eating (and drinking) and may eventually die. Teitelbaum was able to keep alive such animals by feeding them liquid food through a stomach tube; eventually, they will accept very palatable foods of favored taste and smell, but they have no urge to eat equally nutritious foods without these sensory effects. At near-recovery the urge to eat returns, and animals will press a bar to inject liquid food into the stomach even though they have no sensation of taste or smell. In other words, the animal is now adjusting its caloric intake. Drinking is also affected by these medical lesions in a slightly different way, principally in recovering more slowly. The fully recovered rat eats and drinks nearly normally, except that it still rejects food with bitter taste more readily than do normal animals. Thus, with destruction of either the medial or lateral portions of the hypothalamus, the rat becomes more sensitive to slightly noxious food.

In sum, the lateral hypothalamus is a center of feeding motivation that is inhibited by the medial hypothalamus upon food intake. This conclusion can also be verified by several other methods. For instance, localized anesthetization has the same effects as lesions: if the functioning of the lateral hypothalamus is disrupted, the animal fails to eat; if that of the medial portion is disturbed, the animal does not satiate because the inhibitory effect on the lateral hypothalamus is removed. Electrical stimulation has opposite effects: activation of the lateral portion evokes eating, that of the medial portion inhibits eating.

Yet another demonstration of the parameters of the feeding-control mechanism was demonstrated using a self-stimulation technique. Electrodes implanted in the lateral (motivating) hypothalamic areas delivered a shock when the rat pressed a lever. When fed excessively, the rat decreased the rate of the self-administered stimulation, just as it decreased further feeding. That the effect was mediated by the ventromedial hypothalamus was shown by stimulating this area with electrodes and observing a similar decline in self-administered stimulation to the lateral hypothalamus. As expected, ablation or anesthetization of the medial portion increased self-stimulation to the lateral portion and also increased feeding. Thus, lateral hypothalamic self-stimulation is more reinforcing to a hungry rat than to a satiated one so that this effect seems to mimic the gratification obtained by eating.

This brief view of experiments on a classical drive system demonstrates the breadth of experimental techniques that can be brought to bear on central control systems. Many parameters remain to be completely elucidated; for instance, the nature of the sensory input to the system and the nature of the output commands to the effectors of feeding behavior. Furthermore, the coding of messages into and out of the hypothalamus, as well as the coding of the inhibition that the medial satiation portion imposes upon the lateral (motivating) portion need to be specified. The combination of single-unit recording and histological techniques found so useful in sensory coding studies will be increasingly brought to bear on central control mechanisms as well.

HORMONES AND REPRODUCTIVE BEHAVIOR

A ring dove will not always sit on eggs in a nest when they are offered; ordnarily it will do so only during a restricted part of the breeding cycle, when it might be said that a motivational state for incubation occurs. Likewise, many other behavioral responses of reproduction can be elicited only during certain times. Since there are continuing changes in the hormonal state of the animal throughout the breeding cycle, these motivational states are thought to be dependent in part on the hormonal states. Lehrman and his co-workers have elucidated many of the reciprocal interactions between hormonally induced motivational states on the one hand and behavior and external stimulation on the other (e.g., Lehrman, 1958a, 1958b, 1961, 1964; Lehrman et al., 1961; Erickson and Lehrman, 1964).

Normally, both the male and female will incubate eggs as soon as they are laid. If a pair are placed together with a pre-built nest with eggs, they do not incubate; rather, they court, build own nest (on top of the eggs) and finally sit on the eggs some five or more days after introduction. (The eggs have to be continually fished out and placed on top of the new nesting material.) That the five to seven day delay is not due to being in a strange cage was shown by placing birds in the breeding cage for a week but separating male and female by a central opaque partition and denying them both nesting material. When the partition was removed and nest plus eggs provided, the pair still did not incubate for another week.

Does sitting require the sight of the eggs for seven days? This seemed unlikely since normal pairs incubate as soon as the eggs are laid. The point was tested, however, by giving the pair nesting material but no eggs for a week; then eggs (in a new nest) were introduced and the birds incubated immediately. Since single birds, given nest and eggs, never incubate, it appeared that the incubation drive is activated by familiarity with the mate or with nesting material or both for some period. If birds are together for a week and then given nesting material, they do not incubate immediately, nor do they require another week. Rather, they build intensively for a day and then incubate. From these and other experiments it is evident that association with the mate activates a nest-building drive whose expression in actual nest building then activates broodiness.

These behavioral experiments suggested that the motivational states were due to the release of hormones. Similar experiments were then run in which hormones were injected into the doves. Injection of the male gonadal hormone, testosterone, had no effect. Estrogen, the female gonadal hormone, induced intensive nest building for several days, after which the birds incubated. When the ovarian hormone progesterone was administered, however, birds sat on the eggs within a few hours of introduction. Similar combinations of behavioral and hormonal experiments were performed to see what brings about the readiness to feed squabs on the pigeon milk secreted from the crop. Here the hormone prolactin was shown to be important.

If participation in courtship elicits estrogen flow in the female, as indicated by the above experiments, is it necessary that there be physical contact between mates? The answer is no because females who merely observe their mates through glass plates still undergo ovarian development and the behavioral traits associated with it. If the males are castrated, they do not go through the calls and postures of courtship, and in this case the females do not secrete estrogen. Therefore, the specific behavioral patterns of the male bring the female into readiness for nest building and associated activities. Participation in these activities, in turn, leads to a readiness to incubate.

How does progesterone bring about this incubation drive? Since the hormones must somehow affect the nervous system, Komisaruk (1971) tested the possibilities of one such effect. He implanted minute bits of progesterone in the dove's brain, principally in the hypothalamus, and noted whether or not the birds incubated. Histological examination of the brain showed certain spatial locations that reliably induced incubation and others that did not. It may be, therefore, that certain areas are sensitive to hormones and become active when stimulated by them. The next obvious step in this sort of research will be to stimulate these areas electrically and see whether incubation can be induced thereby.

BRAIN STIMULATION

An important series of experiments showing how central nervous activity predisposes an animal to react to certain stimuli was conducted by von Holst and St. Paul (1960, 1963). By inserting electrodes into the brain stem of chickens and recording the behavior of these birds toward various stimuli during the presence and absence of electrical stimulation of the brain, they discovered several interesting interactions between drive states and external stimuli.

Von Holst and St. Paul were able to elicit two broad categories of behavior: simple movements, such as standing, sitting, preening, peering, scratching, and so forth; and complex sequences of movements, such as seeking and eating food, settling on the nest, escaping from predators, and so forth. The simple movements offer little information about drive states, but the complex sequences of behavior interact with external stimuli and the resulting behavior resembles that of normally motivated behavior.

For instance, brain stimulation in certain areas leads to aggressive behavior if an appropriate recipient object is available. A rooster standing next to a stuffed predator will ignore it—until the brain is stimulated, at which time the chicken delivers a violent attack. Stimulation in the absence of the predator, however, elicits little overt behavior, demonstrating that the stimulation does not merely activate a motor center, but, rather, a motivating one. In reality, these areas should not be called centers because few are discrete loci. Rather, they may be tracts, large areas, or more complex topological loci of the brain that represent parts of the underlying "wiring diagrams" of whole behavioral systems.

Von Holst and St Paul also tried stimulating combinations of drives simultaneously. One area, when stimulated, causes stretching in an incubating hen, while another area evokes ruffling. When stimulated simultaneously, stretching takes precedence, suppressing or inhibiting the ruffling. However, after the stimulation ceases, the ruffling appears as a kind of "after-discharge" of motivation. The parallel with behavioral alternation of simultaneously activated drives mentioned above is striking. A more complex phenomenon was obtained by stimulating simultaneously areas that cause sitting and fleeing. A fleeing hen usually looks around, gets up, and finally jumps. When sitting is stimulated however, the hen remains sitting until some threshold is reached and then abruptly jumps away.

Still other typical drive phenomena are mimicked by brain stimulation in chickens. An aggressive peck is given to a stuffed hen or to the trainer's hand, but not *in vacuo*, when a certain area is stimulated. Stronger current at the same brain site, however, does elicit aggressive pecking at the ground, a sort of vacuum activity (cf. above). Brain stimulation also shows the "exhaustive" properties of drive; a drive stimulated repeatedly or for a prolonged time is difficult to elicit a few minutes later, and it may remain refractory for as much as an hour afterward.

Von Holst and St. Paul also found evidence for hierarchies of drives and for displacement activities (cf. above). Modern stereotaxic techniques make possible repetition of these classical experiments with insertion of electrodes in discrete places to allow for specific tests of motivational systems and drive phenomena. Ultimately, brain stimulation studies may allow a relatively complete mapping of the neural systems responsible for evaluating incoming sensory information, both internal and external, for maintaining states of internal readiness and for triggering and co-ordinating complex sequences of motor responses.

CONCLUSIONS

The techniques outlined in this section—brain extirpation, hormone injection and implantation, brain stimulation—along with electrical recording from the brain, are the principal tools by which the actual control systems of complex motivated behavior will become known. The really important trend in such studies is the increasing attention to careful observation of actual behavioral responses and response sequences. The physiology of behavior is the area most likely to command attention of ethologists in the future.

Selected Readings

Pribram, K. 1971. *The Language of the Brain.* Prentice-Hall, Inc., Englewood Cliffs, N.J. (A fascinating account, both factual and speculative, of the relation between brain structure, function, and behavior.)

Roeder, K. D. 1963. *Nerve Cells and Insect Behavior.* Harvard University Press, Cambridge, Mass. (A small book which says much about the study of behavioral control in a group of animals that has proved very rewarding for physiological study.)

Young, J. Z. 1964. *A Model of the Brain.* Oxford University Press, New York. (An attempt to explain how the brain of the octopus works in learning and perception.)

Chapter 12

HOW DOES BEHAVIOR DEVELOP IN THE INDIVIDUAL?

Ontogenetic Interactions

Introduction

Two sorts of differences between behavior patterns have their origins in ontogenetic processes. First, differences may depend on the age of the animal; and second, the behavior of two different animals of any given age may differ. In both cases, the important question concerning the observed differences concerns the nature of the effects the environment has had upon the individual during its development.

Any character of an organism is determined by the interaction between the previous state of the organism and its environment. Even the fertilized egg is not truly determined *genetically* since all but the nucleus is of maternal origin. Some characters may develop similarly no matter in what environment the animal is reared so long as that environment is conducive to life; such behavioral characters are often called *instinctive* or species-characteristic (though some ethologists also imply by this that development is largely independent of environmental influence). However, many (perhaps most) behavior patterns develop differently in different physical and social environments. The most commonly studied processes of interaction are grouped under the rubric *learning*, although other rubrics are also used.

Far too much has been written about learning and instinct. The radiation from such debates has frequently been more in the infrared than in the visible spectrum. There are certainly many patterns of behavioral development. The problem of development is an empirical one and can be solved for any single behavior pattern only by appropriate experiments.

The solution, however, does not lie in apportioning an act to two separate causes, genetic and environmental. The persistence of this dichotomy, as witnessed by the continuing debates on the heritability of IQ, must be a puzzle to disinterested observers. Partly, it may be attributed to the characteristics of our language which encourage, if they do not compel, certain ways of organizing our perceptions (Whorf, 1956). We must organize all our statements about the physical universe into a noun-verb pattern, a pattern that requires a sleight of hand with phenomena that are not manifestly polarized. Thus, "*it* thunders," "the *lightning* flashes," although there is no "it" and the lightning is the flash and not a separate actor. I suggest that the nature-nurture controversy is at least partially the result of a language ill-designed to deal with processes that are not subject to an algebraic analysis.

Another contribution to the nature-nurture controversy has resulted from a faulty conception of what is meant by "genetic determination."

When behavior is labeled "instinctive," we assume a highly constrained relationship between an act and the structures that subserve it. Those structures, in turn, are assumed to be similarly linked to specific regions of particular chromosomes. The link between specific chromosomal loci and particular physical attributes may indeed be a closely coupled one. Such coupled attributes are commonly said to be "genetically based" or "heritable" or even "innate." These expressions, however, must not be allowed to imply that the information for this expression of the attribute is wholly contained within the chromosomes. The genetic code, far from being a blueprint of any given organ or organism, is an information generating device whose effectiveness depends on its being allowed to function in an environment that already contains a high degree of order.

Consider the preference of a gull chick for a particular color. Color preference in gulls depends, at least partially, on the presence and distribution of oil droplets in the retinal cells (Hailman, 1967). How is their distribution and appearance controlled by the genes? The simplest assumption is that at a particular locus of a chromosome, a cyclical process is initiated which leads to the production of RNA, which reacts with *extra*-chromosomal substances to produce other substances that derepress or repress other chromosomal loci, leading, eventually and after numerous feedback cycles, to the formation of certain oil-soluble pigments. Obviously, this description is but a small distance of the way to a color-coding and perceiving mechanism. But the question of what in fact has been inherited becomes even more difficult to answer when we recognize the complexities of the mechanisms that underline color preferences. In such a cybernetic system there is no meaning to a chromosomal-nonchromosomal assignment of origins.

Studies of heritable differences in behavior would become more intelligible if we were to view the gene not as a repository of data or a blueprint from which an organism can be constructed (that is, as an inchoate homunculus) but rather as an information generating device which exploits the predictable and ordered nature of its environment (or in Schroedinger's terms, 1951, which feeds on negative entropy). This view accords well with current models of gene action, such as that advanced by Jacob and Monod (1961), among others, or that of Waddington (1966).

Just as the phylogenetic history of behavior is due to the action of natural selection through time, the ontogenetic history is due to the action of and changes in the control mechanisms (see Chap. 11) through time. We know little as yet about such interactions before hatching or birth. These early ontogenetic processes may differ considerably from common types of learning; they are akin to the yet mysterious embryonic processes of differentiation (Hamburgh, 1971). Whether *post*-hatching and *post*-birth learning processes are merely special cases of embryonic differentiation remains to be proved; Thorpe (1963b) believes they are fundamentally different.

Variations in the Parental Endowment

Differences in behavior between two animals can be attributed to differences in the parental endowments if the animals have been raised in identical environments. Such differences are often called *genetic*, but they are not necessarily so. For instance, the difference could be due to the maternal cytoplasmic endowment to the ovum, in which case the difference might be said to be due to nonnuclear inheritance. Note, however, that to speak of the heritability of differences in behavior is not the same as speaking of the heritability of behavior.

Furthermore, the behavior of the mother during pregnancy can affect the development of the embryo (example below). Even though such a case may ultimately be due to (*in utero*) environmental influences on the developing animal, the variable observed (experience of the mother) does not act directly on the embryo so that this appears experimentally as an effect of the parental endowment.

In practice, the difficulties of ruling out the possibilities of cytoplasmic or *in utero* influences have been ignored and correlations have been sought between known differences in the parents and differences in the behavior of the offspring (i.e., classical genetic studies). It should be noted, however, that neither the pre-birth nor post-birth environments can ever be completely equated, or even controlled, in genetic experiments. One can only attempt to equate or control those environmental influences that seem to be relevant.

GENETIC VARIATIONS

Few, if any, behavior patterns segregate in classical Mendelian ratios when breeding experiments are performed. The principal reason for this would seem to be that functional behavioral patterns are usually the product of activity in a complex neuromuscular system, which must be highly resistant to chance mutations. Because of their complexities, the behavioral mechanisms are likely to be affected by many genetic loci so that changes at any single locus *would not* produce marked changes in the functioning of the entire system. Also, to be resistant to chance mutations, the neuromuscular system probably contains much redundancy and many compensatory mechanisms so that small genetic changes *could not* produce gross alterations in the functional behavior. We do not expect, then, genetic changes to produce qualitative alterations in behavior; rather, we should expect that behavior is polygenically controlled and that breeding experiments will show quantitative changes in behavior. The converse of polygenic effects (many genes affecting one character) may also be expected to occur as pleiotropism (one gene affecting many characters), in which case some of the overt behavior need not be functional (see Chap. 9).

The first extensive attempts at analyzing the genetics of behavior were attempted by Tryon in his studies of maze learning in rats (see Chap. 3). Much of modern behavioral genetics has also concerned the inheritance of differences in abilities (see below). A more discrete genetic effect was discovered by Bastock (1956), in which a single gene mutation leads to yellow-eyed *Drosophila melanogaster* fruit flies. The mutants have reduced success in mating with normal females. Certain observations and experiments exclude the possibilities that the females react against the yellow color or against some altered scent. Instead, there is a linked change in the wing display (vibration) of the yellow mutant males that causes vibration to occur in shorter bouts and at greater intervals. However, there are few comparable examples, and the genetic basis of most behavioral differences seems to be multifactorial.

If the genetic differences are great enough, qualitative differences in behavior can sometimes be studied genetically. Dilger's (1960) studies of nestbuilding in lovebirds provides another more recent example. One species of lovebird carries nesting materials by tucking several strips of grass into the feathers of its rump. A closely related species carries single strips to the nest site in its beak, in the more usual fashion of birds. The interspecies hybrids try to place strips in their rump feathers but are rarely successful at first. But they persist, and apparently with long practice they may be successful carrying at least a few strips in the manner of one of the parent species.

Many similar studies have been done on interspecies hybrids in birds, and some of these are reviewed by Hinde (1956*a*). His results from hybrids among two wild and one domestic species of cardueline finches are typical of this sort of experiment. All the behavior that is identical in the two parental species appears unchanged in the hybrids. If the behavior of the two parental species differs quantitatively (form, duration, frequency, etc.), the hybrids exhibit intermediate behavior. If the behavior of the two parent species differs qualitatively but is apparently derived from intention movements shared by both parent species, a somewhat similar result obtains. For instance, the European goldfinch has a "pivoting" display, presumably derived from locomotor movements shared by the other species, although the other species do not have the derived pivoting display. Hybrids showed a movement intermediate between pivoting and the unritualized locomotor movements (which they also possessed). These general results strongly suggest a polygenic basis to the differences in the behavioral patterns studied.

Some exceptions to the above results have been reported elsewhere. Certain pheasant species differ in foraging habits. Huxley (1941) reports that a hybrid between one species that scratches for its food and one that digs for it with the beak resembled the latter parent. Examples such as the pheasants and lovebirds—in which the F_1 hybrid's behavior is more like one parent than like an intermediate between the parent species—seem to involve behavior patterns that normally develop gradually. That is, no clear dominance is seen in the naïve hybrid, but it can learn a particular response and the learned response resembles more the response of one parent than the other.

Heinroth, and later Lorenz (e.g., 1941), described the behavior of hybrid waterfowl. Usually, the hybrid's behavior is intermediate between the parents, but sometimes the hybrid resembles one parent somewhat more than another in characters like "wildness." Similarly, Hinde (1956a) found evidence that hybrids between domestic canaries and other finches tended to be nearly as wild as the wild parent. When the genetic relationship is distant but hybrids can still be obtained, the offspring often show no displays and related social-sexual behavior at all, even under hormone administration (Poulson, 1950). Sometimes a sterile avian hybrid will nonetheless show incubation and parental behavior if provided the correct situation (Mainardi, 1958) even though they cannot themselves reproduce, thus demonstrating the genetic independence (in this case, in pigeons) of sexual and parental behavior.

Insects, of course, provide better material for genetic experiments because so many can be bred so rapidly in the laboratory. Haskins and Haskins (1958) studied the behavior of cross-species hybrids of two cocoon-spinning moths. First instar larvae resembled one parent species, *Callosamia angulifera*, and made the same food choice as this parent so that food preference appears to be completely dominant. However, cocoon-spinning showed all grades between the parents, thus indicating a polygenic inheritance.

Drosophila fruit flies have been the commonest experimental animals for genetic studies of behavior, and a few results of these studies may be mentioned to indicate the kinds of genetic effects to be expected. Manning (1961) selected for both fast and slow mating speeds for 25 generations of *D. melanogaster*. Selection was only really effective over the first seven generations. F_1 hybrids showed an intermediate mating speed. The "slow-mating" flies showed greater general activity than did the fast ones, but "fast" and normal control flies had a higher intensity courtship. Thus, the artificial selection worked, not on general activity, but primarily on differences in nervous thresholds specific to sexual behavior.

In a somewhat similar experiment, Ewing (1961) selected for large and small males in the same species. Thus selected, small flies did *not* resemble some other, naturally occurring small flies created by environmental effects on development. In the selected lines, small males possessed increased vibration during courtship and the large males decreased vibration. The difference was not due to direct correlation between size and courtship vibrations since in F_2 hybrids there was no correlation between these two characters. Further, no correlations between general activity and body size appeared, although both selected lines of flies showed somewhat lower general activity than did normal flies. It turns out that when placed in competition for mates, the small males are less successful than the large ones. To test whether the increased vibration had occurred as "compensation" for small size, small flies and large flies were reared for several generations with no competition between them for females. A decrease in vibration intensity did occur in the large flies, suggesting that the change in behavior was not due to competition during courtship. On the other hand, the small flies did not develop increased vibration—until large flies were introduced with which they had to compete!

Scott and Fuller (1965) attempted a thorough analysis of breed differences in domestic dogs. They demonstrated that breeds of dogs differ significantly in their performance in a variety of behavioral tests, including both tests of emotionality and of different kinds of learning ability. Most significantly, they found relatively few traits for which any breed is actually homozygous. Even with restricted samples of inbred lines, genetic variability remained high. This reinforces the view expressed above that most (though not all) differences in behavior are the outcome of multiple interactions rather than the expression of single genes.

One other important finding relates to the rates of age-dependent changes in behavior. These show significant breed differences. For instance, in one series of tests, wire-haired terriers showed an increase in emotionality scores as they aged from 17 to 51 weeks, while Shetland sheep dogs decreased their scores over the same period. Both breeds had initially had similar scores. This result underscores the importance of considering a temporal dimension along with the genetic and environmental interactions. A given environmental stress applied at one age might provoke identical responses in shelties and terriers, while at a later age the responses would differ.

Marler (1963) reviews some recent evidence of heritability of differences in sound behavior of animals. In certain Orthopteran insects (e.g.,

Nemobius and Chortippus) F_1 interspecies hybrids have songs intermediate between the parental types. An exception seems to be *Gryllus*, in which two parental species differ strikingly: one sings before copulation and the other does not. Here genetic analysis of F_1, F_2, and backcross individuals indicates a monofactorial inheritance. In anuran amphibians (principally the work of Blair and Bogert), F_1 interspecies hybrids within the genera *Hyla, Microhyla, Bufo*, and *Scaphiopus* all give calls intermediate between the parental types, thus providing further evidence for the generality of multifactorial inheritance of behavior.

NONGENETIC VARIATIONS

Darwin, as related earlier (Chap. 1), believed in the possibility of learned traits being inherited. This view still found favor as late as the 1930's, as evidenced by McDougall's (1938) experiment on the heritability of maze knowledge in rats. In the decades following, as an understanding of gene action grew, the assumptions implicit in such studies as those of McDougall were generally rejected.

Since 1950, however, the possibility of nongenetic heritability has been broached in other forms: first, heritability of emotional states through mediation of the endocrine system; second, inheritance of learned behavior by ingestion of specific nuclear acid fractions.

Evidence for the nongenetic transmission of affect comes from a study by Keeley (1962) (see also Denenberg and Whimbey, 1963). Keeley subjected albino mice to stress by crowding, a condition known to alter adrenal function and behavior (see Christian, 1963, for a detailed review of this latter topic). The offspring of crowded and uncrowded mothers differed in activity and "emotionality" (as measured by defecation rate and response latency), irrespective of whether rearing was by their own or a foster mother. That is, the difference in the youngsters' behavior was solely a function of the stress suffered by their mothers during pregnancy. Keeley suggested that "aberrant endocrine activity in the crowded, pregnant female impairs the development of fetal response systems" (p. 44). This work has been corroborated in other labs (Denenberg, 1964).

The evidence concerning nongenetic transmission of learned behavior is considerably more controversial. Originally, it was claimed that in certain flatworms (*Planaria*) the ingestion of fragments of a "trained" worm by another enhanced the learning rate of the cannibal. Treatment of the fragments with enzymes capable of denaturing RNA abolished the effect (see McConnell, 1964). Subsequently, it has been claimed that RNA taken from the brains of trained rats and injected interperitoneally into naïve rats leads to a partial transmission of the learned trait (Babich et al., 1965).

Whether these and similar experiments reported in recent literature can be replicated in other laboratories and how they are to be interpreted cannot be foretold at this moment (see John, 1967; Deutsch, 1969; and Glassman, 1969).

Variations in the Environment

Great interest surrounds the problem of which kinds of experience influence behavioral development and how these experiences bring about its effects. For convenience, we discuss some recent studies under four separate headings; the concept of critical periods during behavioral development, effects on perceptual and sensory phenomena, effects on motor activity and co-ordination, and effects of experience on behavioral organization. The last category is really a potpourri that includes classical drive systems, affectional systems, and similar complex forms of behavior that are not easily classified as sensory or motor.

CRITICAL PERIODS

It has been generally accepted that many organisms are more capable of learning particular tasks at certain ages or developmental stages than at others. In ducks or geese, the tendency to approach a particular model of a series of differently colored or shaped ones may be contingent upon exposure to that model within a relatively short period following hatching (see Sluckin, 1964, for a detailed discussion and bibliography on this topic, as well as the other aspects of imprinting discussed below). In humans, Penfield and Roberts (1959) have suggested that language learning occurs most readily before a certain age, or at least that it involves different neural mechanisms at later ages. In general, the limits of these optimal or sensitive periods for the learning of particular tasks or discriminations are not sharply defined.

Occasionally, however, instances do appear where such optimal periods follow a time course that resembles that of the embryologists' *period of competence* (Weiss, 1939). For example, Ramsay and Hess (1954) found that in order to develop a given degree of preference for a particular model, their mallard ducklings had to be exposed to that model within a span of approximately 10 hours. When the optimal period for the development of such preferences is relatively brief, or its temporal limits sharply defined, the term *critical period* has conventionally been applied. The important question which we ask here is, given a critical period, what are the proximate determiners of its onset and terminus? In particular, to what extent is the timing a function of sensory maturation?

It is axiomatic that visual responses cannot be learned prior to myelinization of the optic tract and the establishment of functional synapses in the nervous system. But, whether it is this kind of maturational event that determines the onset of a critical period it is impossible to say. Part of the difficulty in providing an answer stems from the paucity of examples of true critical periods. If we could be certain, for instance, that the periods of socialization in dogs described by Scott (1962) were critical, it would be useful to study the changes in sensory and perceptual capacities of dogs at and before these periods. Neither has the critical character of the developmental periods been confirmed, nor changes in the dogs' sensory system examined. Obviously, there is little point in seeking the latter until we are certain of the existence of the former.

In ducks, where the existence of a critical period for the elicitation of the following response seems reasonably clear (at least under certain conditions of rearing; see Sluckin, 1965), a somewhat more hopeful experimental situation exists. Klopfer and Gottlieb (1962a and b) suggested that in mallard ducklings, the critical period for the development of preferences for auditory stimuli may precede that for the development of stimuli visually perceived. If critical periods are initiated by changes in sensory function, it should be possible to relate age and sensory capacities in these ducklings. To date, this has not been undertaken (but note Gottlieb, 1970a and b).

Most of the theories created to explain the terminations of the critical period do not invoke changes in sensory capacity *per se*, though such changes are generally implied. Two general categories of explanation that have been offered can be summarized thusly (see also Kaufman and Hinde, 1961):

1. Changes in fearfulness and anxiety: Moltz (1963) has been the principal champion of the view that changes in the anxiety level determine the end of the critical period for certain functions (e.g., the following response of ducklings or chicks). Whatever the experimental data, and these conflict on the question of whether chicks are more or less fearful after the critical period, the alleged explanation begs the question. What underlies these motivational changes? Are these the consequence of altered sensitivity or filtering of input (sensory) channels and transducers or changed central organization? Are the changes endogenous, i.e., independent of information from the external world, or are they shaped by external events? The finding that alterations in the brooding conditions may influence the

duration of the critical period in ducks and chicks indicates that whatever the role of developmental changes, these are at least influenced by experience.

2. Changes in the level of the sensory input: Suppose we consider the critical period to be an evolutionary strategem that assures that the individual will be responsive to and thus learn stimuli that are presented to it at a particular time. The support for this view has been summarized by Thorpe (1963a). Then it is reasonable to assume that once the critical period has begun, the organism will remain responsive until the learning has occurred. The signal for "learning completed" could be a certain level of sensory loading, i.e., a particular kind or degree of stimulation. Thus, chicks reared under conditions of visual deprivation should have much longer critical periods than those reared in light or with siblings. This is essentially the argument that has been advanced by Sluckin and Salzen (1961). The data thus far available are compatible with this view, though the paucity of undoubted instances of critical periods makes it difficult to devise tests.

One instance of a critical period whose explanation fits neither of the above two categories is known. Parturient goats will generally accept and care for their young only if they are in contact with them during the first few moments after birth. The contact need not extend over more than five minutes to lead to the formation of a stable maternal-filial bond, nor can a later contact, e.g., an hour following birth, remedy the effects of an immediate post-parturitive separation (Collias, 1956; Klopfer, Adams, and Klopfer, 1964). Apparently, the dilatation of the cervix that attends birth causes a release of oxytocin, which, in turn, is involved in "turning on" the maternal-affection system (Klopfer, 1971). The cues by which the mother recognizes her young appear to be olfactory (Klopfer and Gamble, 1967).

In this example, the "critical period" is not tied to a specific age of the animal, but, rather, to a specific point in the reproductive cycle, a point which may occur many times during the life of the individual. What relation this latter sort of "critical period" bears with critical periods in imprinting (above) and avian song learning (below) has yet to be made clear.

In short, the readiness of the mother to accept her baby wanes extremely rapidly. Observations suggest that neither anxiety states nor stimulus loading are involved in the rapid decline in the individual's readiness to accept young. Perhaps we are here dealing with rapidly decaying hormonal effects (for instance, Folley and Knaggs, 1965). Alternatively, there may be highly labile and attractive (to parturient females, only) substances associated with the birth fluids. As these decay, their attractiveness wanes, as does the critical period. Thus, critical periods of different animals and subserving different functions may be regulated in quite different fashions.

Of interest here, too, are some experiments performed by Vince (1960) concerning the age dependence of simple investigatory responses in great tits. Birds are trained to remove the white lid of a dish containing food. They are then tested with a successive series of food dishes with white covers and empty dishes with black covers, presented in random sequences. The test was to measure how well the birds could refrain from approaching the black dishes. Young birds respond immediately (within 10–15 seconds) to the black lid; latencies rise to 15–20 seconds by 25 weeks after leaving the nest, and to 45 seconds by 35 weeks. However, at 70 weeks of age, birds again respond rapidly. One interpretation could be that the ability to discriminate rises (then falls) with age, but a second experiment renders such a simple hypothesis doubtful. Tits given merely an object to respond to will peck and pull at it for differing periods of time as a function of age. In this case, however, there is an early peak of positive responsiveness (about 15 weeks), dropping sharply to a low level by 30–40 weeks.

Part of the explanation of the seeming age dependence of discrimination learning may thus be that young birds simply cannot refrain from exploring a novel object. Therefore, though they may be able to discriminate the white and black lids, they continue to pull at the unreinforced black ones until reaching an age when the investigatory urge has subsided. Perhaps the waning of this urge involves a sort of sensory loading (above). Still unsatisfactorily explained, however, is the increased responding to the black lids in older birds (70 weeks) at an age when the investigatory response (as measured) is very low. This example shows how complex the subject of critical learning periods may be and how misleading it may be to study one response in isolation from others.

EFFECTS ON PERCEPTION AND PREFERENCES

In some cases, classical forms of conditioning or other learning seem to be responsible for ontogenetic changes in perceptual preferences, but in other cases the role of experience is less well-understood. As no general principles seem yet to have emerged, it is valuable to review a few of the better studied examples to show the variation and importance of experience in animal perception. One such example concerns the phenomenon of imprinting.

Imprinting is generally regarded as the establishment of a stable perceptual preference as a consequence of a previous exposure to the preferred item during a short-lived critical period. The most generally accepted criterion for the occurrence of imprinting, at least in the precocial ducks and fowl which have most commonly served as subjects for imprinting studies, is the elicitation of the following response. Most precocial birds will, at an early age, follow slowly moving objects that precede them. At a later age the following response disappears, unless it has been previously elicited. If the following response is still elicited by a particular model to which exposure was made during a critical period after this period has passed, then it is believed that this behavior serves as a valid criterion for the occurrence of imprinting. Alternatively, if the imprinted object—that is, the object presented during the critical period—is preferred to other objects in a simple choice situation, then this, too, is taken as a criterion for the occurrence of imprinting, i.e., the existence of a stable perceptual preference.

There are, unfortunately, methodological objects to this procedure, common as it has been. First of all, Schutz (1965) has argued that a choice of following surrogates made at an age of generally no more than several days may bear no relation to other preferences manifested later in life, particularly the sexual preferences, which also appear to be established at about the same age. The economic problems of raising water fowl (or other animals) to maturity have generally prevented people from accepting Schutz's admonition, i.e., re-testing preferences at maturity. A few workers have attempted this, notably Schein (1963). Alternatively, some investigators have sought to induce sexual precocity with exogenous hormones in the thought that the choice of a copulation partner made by a sexually precocial youngster would give an indication of imprinted sex preferences. One may, however, doubt whether precocial sexuality leads to similar object preferences as does normal development.

The second major objection to the use of the following response as the criterion for the occurrence of imprinting is to be found in work by Klopfer (1964) and Klopfer and Hailman (1964). They found that while the imprinting experience, that is, the initial exposure of the subject to a surrogate during its critical period, was in fact essential for the subsequent development of certain preferences as measured by the following response, the character of the preferences was not always influenced by the original surrogate. For example, if chicks or ducklings were allowed to follow one of two models (plain and colored) during their critical period, and later exposed to both models, the results varied according to whether the models were stationary or moving. Under stationary conditions, the birds approached their imprinted surrogate. With the models moving, the colored model was more strongly selected, regardless of which model the birds were originally exposed to. Controls, not exposed during their critical period, evinced no preferences. Clearly, exposure to one or the other model during the critical period was a *sine qua non* for the development of a preference, though the character of the preference is not immediately obvious. It may be that the cues by which the model was recognized varied according to whether the models moved or not. Perhaps some feature such as visual flicker predominates during movement, and color or shape at times of rest.

Aside from the methodological considerations cited above, recent evidence requires a further qualification of the classical notion of imprinting, i.e., the notion that the organism is initially a *tabula rasa* onto which a preference is imprinted. This evidence demonstrates the existence of preferences that favor imprinting upon certain stimulus patterns rather than others. Gottlieb (1965) found that naïve ducklings and chicks prefer the parental calls of their own species to those of others. Imprinting to calls of alien species may be possible but not in the presence of the "correct" specific call. An earlier study of Gottlieb (1961) can also be interpreted as demonstrating a change in the range of acceptable imprinting surrogates (visually perceived) as a consequence of domestication. Thus, perceptual mechanisms may increase the probability of a particular experience, as compared to some other experience, resulting in imprinting (see also habitat selection, below).

Another example of sensory changes in ontogeny is afforded by gull chicks (Hailman, 1964a). The chick pecks at the parent's beak in order to beg food. The newly hatched chick pecks most readily at objects with certain simple characteristics: an elongate shape with vertical orientation, a certain width, red color and which is moved horizontally across its long axis (see Chap. 11). Week-old chicks taken from the nest, however, will peck only at a model which closely resembles the head of the parent with whom they have had feeding experience.

For instance, newly hatched chicks of two species (herring gull and American laughing gull) show no preference for their own species, whereas week-old, experienced chicks prefer their own species. That these developmental changes in form preference are due to conditioning was shown by an experiment in which chicks were rewarded for food when they pecked at a model of the wrong species; they came to prefer this species. Therefore, species recognition, at least in this behavioral situation, is learned through conditioning. Interestingly, no striking changes could be shown in color preference, and chicks seemed particularly resistant to conditioning to a different color. Thus, when the preference at hatching already matches closely the functional stimulus (e.g., red color), experience has little effect; when the initial preference is more general (e.g., form preference), rapid conditioning takes place.

Finally, we may consider the development of habitat preferences. Animals restrict the bulk of their activities to a particular portion of the available environment. The ecological implications of this fact are not relevant in the present context (see Klopfer and Hailman, 1965), but the behavioral implications are. How do animals recognize the appropriate habitat? How are preferences for particular habitats established? They may, of course, be partially imposed by morphological constraints in the manner that a webbed foot favors swimming over hopping. They may also result from imprinting or conditioning. The problem has been to specify the components of the environment and then to determine when and how the preferences are generated by comparing animals reared under differing degrees of isolation.

Two studies will illustrate this approach. One by Wecker (1963) compared under natural conditions the habitat preferences of two strains of deer mouse, one wild-trapped, the other from lab-stock, reared in either the lab (naïve animals) or woods or fields. His results showed that the choice of fields by the normally field-dwelling strain is reinforced by early experience in fields, but such experience is not essential. Experience in woods or lab does not reverse the normal affinity for fields. The lab strain, however, showed a markedly reduced affinity for fields; early exposure to fields was necessary for a strong field preference to develop. Exposure to woods, on the other hand, did not lead to a corresponding preference for the woods habitat. Wecker concluded that a form of habitat imprinting was operative in these mice, but a perceptual predisposition existed that led to the favoring of one habitat over the other.

The second study, by Klopfer (1965) and Klopfer and Hailman (1965), sought to determine what components of their habitat were recognized by birds and then to rear birds under different conditions in the manner of Wecker. Leaf size, shape and light-shadow patterns appeared to be of importance to at least some birds. Further, some preferences were more easily established than others, just as in the case of Wecker's mice. Thus, habitat selection, like other forms of important perceptual behavior, is the outcome of interactive processes.

EFFECTS ON MOTOR ACTIVITY AND CO-ORDINATION

The fact that a complex stereotyped motor pattern is exhibited by all members of a sex or species does not prove it is "unlearned." (After all, nearly every American boy swings a baseball bat in a certain manner, yet the development of the motor pattern clearly involves learning common to the whole population.) One example of this is the motor pattern of pecking shown by begging gull chicks (Hailman, 1964a). It involves three principal actions: opening of the mandibles (and subsequent closing around the parent's beak), movement of the head forward (and return), and rotation of the head so that the chick's mandibles can grasp the vertically held bill of the parent. All week-old chicks show the full pattern, but newly hatched chicks lack the rotational component and as a result usually fail to grasp the parent's beak, striking it with their mandibles instead. Chicks reared and force fed in the dark to an age at which wild-reared chicks show the rotation also fail to demonstrate this component. Just how this rotational component develops as a result of feeding experience is not yet known.

Another aspect of the motor side of the gull chick's pecking response is the accuracy with which the peck hits the directing stimulus. Newly hatched chicks strike a model of the parent's bill only about 50% of the time. Accuracy increases rapidly with age, reaching asymptote in just a few days, at 90 to 95% hits. In this case, accuracy improves even in the absence of practice (i.e., in dark-reared chicks) and is probably due to general motor development (postural strength, etc.). The initial inaccuracy turns out to benefit the chick, for in aiming at and missing its parent's or experienced sibling's bill, it often strikes and discovers for the first time the food, which it does not recognize visually when newly hatched. Similar postural facilitation of accuracy in pecking has been discovered in domestic chicks. In this species, force feeding during dark rearing can cause the development of atypical motor patterns not seen in ordinary chicks, a result that again emphasizes the important role of experience in normal motor development (see literature cited in Hailman, 1964a).

Perhaps the model experiment for studying the development of behavior is the study of bird song. Here, various degrees of auditory experience can be manipulated, and even atypical experience experimentally programmed into development. The results of these studies have shown a bewildering array of developmental patterns.

Complete acoustic isolation can be obtained by deafening the young bird shortly after hatching. Konishi (1963), for instance, found that all sounds made by the domestic fowl develop normally in deafened birds. Deafened European blackbirds still sing many phrases similarly to normal birds (Messmer and Messmer, 1956). In many species, however, birds reared in acoustic chambers, or deafened, sing amorphous phrases and bits of song, never developing the full pattern, or they develop quite abnormal lengthy songs. Often individuals of these species (if not deafened) can be made to alter their song by playing specific sounds to them during development (Thorpe, 1961, reviews some of the main literature; Marler, 1963, and Nottebohm, 1970, make a penetrating synthesis of avian song learning). Nicolai's (1964) extensive study of song in African finches has shown that these nest parasites learn completely the song patterning of their host through imprintinglike processes. The role of experience in song development thus spans the spectrum from maximum to minimum.

An illuminating intermediate case is provided by the chaffinch (Thorpe, 1961). The normal chaffinch song is a tripartite affair, the terminal flourish of which varies considerably; one bird may have several songs differing largely in the terminus. The general structure of the repetitive notes comprising the song develops somewhat similarly, regardless of the rearing experience; their frequency range and temporal patterning are relatively constant within limits, although the details of these features change depending on the bird's experience. The perfection of these notes and the division of the song into its normal parts are dependent on hearing the normal song prior to the bird's first September. That is, these features normally develop through hearing adult males (primarily the father) sing during the young bird's first few weeks of life. There is no question of immediate imitation here since the young bird does not actually sing its song until the following spring so that some kind of imprinting or latent learning is involved.

Finally, the terminal flourish and certain other details develop in the first spring when the young bird has taken up its territory. Here, it directly imitates the songs of neighboring males which it encounters in singing duels. These details, however, are not relearned with every spring but are established for life during this critical period of the first spring.

We should not suppose, however, that the motor responses involved in song or other vocalizations are representative of motor patterns in general. (For instance, there is a negative correlation in the development of vocal and general motor abilities in children.) Studies of song learning, in short, have raised a variety of important questions involving the role of critical periods, mimesis, and self-reinforcement (Nottebohm, 1970).

Finally, the adaptive responses shown by gull chicks on a visual cliff vary with experience. Emlen (1963) showed that newly hatched herring gulls will withdraw from the edge of an artificial cliff, an adaptive response preventing injury from falling. Hailman (1965) found the same response in newly hatched chicks of the cliff-nesting Galapagos swallow-tailed gull and the surface nesting laughing gull. With older chicks, however, Emlen (op. cit.) found that their behavior was a function of their previous experience. Chicks reared on flat ground behaved as newly hatched chicks, withdrawing from the cliff. Chicks from hillside nests, however, were content to sit next to the edge or to jump from the cliff. Emlen further showed that the difference between these motor responses was due to experience by switching eggs from hill nests with those from flat-ground nests. This example probably does not demonstrate motor development *per se* but reflects a change in willingness (or lack of fear) in jumping over the small cliff.

EFFECTS ON BEHAVIORAL ORGANIZATION

There are many studies of behavioral development that deal with phenomena broader than merely change of the preferred stimulus situation or change in activity or motor co-ordination. Instead, these studies deal with the development of broader, motivational aspects of behavior or with the development of general abilities to learn or tendencies to respond that may be manifest in many different ways.

Although a particular experience may affect only a fragment of an animal's behavior, it can also have more far-reaching, all-inclusive effects. A lab-reared duckling exposed to a moving model at a particular age will, for some weeks, prefer that model to others. There may be no other difference between it and other wild-reared ducklings that prefer their own mother to any model. On the other hand, if a newborn goat kid is not presented to its mother within five minutes or so of birth, it may be gently repulsed when it first seeks to nurse. Even a gentle repulsion may make the kid more hesitant when it next seeks the udder, which, in turn, appears to stimulate the mother to a more severe repulsion. A positive feedback develops, culminating in a total and permanent rejection of the kid by the doe, with the end result being an animal lacking much of this species' usual social behavior (Klopfer and Gamble, 1967).

Several instances are now known in which a fairly specific environmental deficit or stress promotes a major behavioral reorganization. Best known of these are the studies of rhesus monkeys by Harlow (1965). By rearing young monkeys under a variety of conditions, Harlow found that the absence of "mother love" led to severe emotional disturbances, even though the animals' physical needs were met. At first, Harlow believed that the tactile comfort provided by a terry cloth-covered surrogate was all that was needed to compensate for the absence of a mother. Later, he found the "terry cloth-reared" animals to be incapable of forming normal social attachments or of child rearing. The precise role of the mother in preventing the appearance of the "surrogate syndrome" is not altogether clear. It is possible that the mother is responsible for teaching far more of the species-characteristic behavior than has been generally realized. Thus, the disturbances that appear in motherless animals may result from lack of appropriate learning opportunities as much as from emotional or affectional states. For instance, Goodall's observations (film shown at International Ethological Conference, Zurich, 1965) of wild chimpanzees show that (1) a pre-puberal female is eager to hold an infant, (2) is at first unable to hold it comfortably, with the result (3) that the infant squawks and is promptly retrieved by its mother. This sequence is repeated, but on succeeding trials the interval between (1) and (3) lengthens, i.e., the young female becomes, with practice, more adept at handling the infant. The suggestion is that the imitation and practice afforded by the presence of a normal family relationship provides an essential learning opportunity. Without this opportunity, the young chimp cannot later develop maternal behavior.

Studies of rodents, such as those by Denenberg (1964) or King (1958), suggest that in these animals, too, experiences that initially merely alter thresholds for aggression or fear may have numerous and multiplicative effects upon all of the adult social and maternal behavior patterns.

Finally, we should indicate that not all early experiences that affect the organization of behavior necessarily have irrevocable effects. All indications from Harlow's studies are that the behavioral disturbances shown by his monkeys are relatively permanent. However, in ducks, Waller and Waller (1963) have shown that the asocial behavior of isolate-reared ducklings can eventually be altered. By compelling their isolates to associate through confinement in a small pen, they believed they were able to reconstitute normal social behavior. It is a matter of no small importance to clinicians to discover why early experiences may in some instances produce disturbances resistant to treatment, while in other instances recovery may be possible. Such variables as the duration and timing of the trauma (or deficit) require considerably more systematic study.

The ontogenetic organization of responses into functional and motivational systems has received little detailed attention thus far, although two examples can be cited. First, the newly hatched gull chick pecks at the parent bill and at food with the same movement (Hailman, 1964*a*). It is not possible to classify the behavior rigidly into either a begging (communicative) function or a feeding (ingestive) function. In fact, the chick does not recognize food visually and must learn it by trial and error, either by missing the parent's bill at which the peck is aimed (above), by random pecking or by pecking at and missing the bill tip of an older sibling that is feeding (see Social Interactions).

Once the chick does learn to recognize food, however, the motor patterns of begging and feeding begin to differentiate from the initial pattern. The begging peck acquires a rotational component that allows the chick to grasp and stroke the length of the parent's vertically held bill (above). The initial peck also differentiates into a feeding movement of striking hard into, firmly grasping, and tugging at a food mass.

The motivation of both these functional behavior patterns is hunger, which is known to increase both the begging and feeding frequency in older chicks. The question may be asked whether the undifferentiated peck of newly hatched chicks is also affected by the state of food deprivation, and if so, whether the effect requires experience for its development. Only a preliminary answer is available. Chicks reared in the dark so that they did not peck were force fed by placing food well into the throat. At a few days of age the chicks were fed to satiation, and then at various intervals afterward (i.e., at various stages of food deprivation) brought into the light and their pecking rates to a model parent assessed. Increased hunger caused increased peck rates, even though these chicks had never pecked before. It is possible that the food deprivation had a general systemic effect on the chick's physiology that brought about greater activity of all motor patterns and was not specific to pecking. Even so, a motivational basis for further development through conditioning is available to the newly hatched chick and hunger is not a completely learned drive for the pecking system.

A similar type of organization for behavior of the domestic chick is suggested by the studies of Andrew (1963, 1964a, 1964b). A newly hatched domestic chick will, when confronted with a stationary three-dimensional object of about its own size, direct a number of undifferentiated behavioral responses toward it, including approach, exploratory pecks, rapidly repeated pecks, pressing of the body against the object, and climbing on top of it. In response to a broad category of auditory and other stimuli, the chick will also give a repertory of calls. Injection of testosterone into the chicks brings about a number of changes in these generalized responses so that the motor and vocal behavior patterns resemble those of adults (aggression, copulation, etc.) but are not given in the specialized social contexts in which the functional behavior appears in adults. It is suggested that the role of experience is to organize these patterns into motivational categories through conditioning since in normal development any given pattern can only be expressed in certain contexts in which the correct eliciting stimuli are present and the correct reinforcing stimuli ensue. It seems likely that this sort of approach to the study of ontogeny of behavioral organization will become increasingly common.

Social Interactions and Experience

A truly comparative approach to behavior must include studies of man. It has been a fundamental tenet of anthropology that much of the behavior of man is culturally determined, so that behavior is cultural specific rather than species specific. Culturally specific behavior is most likely to be established through learning processes that are channeled by the social environment in which the organism develops, although there may also be genetic differences between isolated (culturally distinct) populations. It therefore behooves the ethologist to study traditions, animal cultures, and social effects on learning, both for their intrinsic value in understanding the behavior of specific animals and for the possibilities of uncovering explanations of behavior that might also apply to man, although, as noted before, not by direct extrapolation.

In the previous section we noted that the behavior of a conspecific (e.g., the mother) can channel the behavioral development of another (e.g., the offspring). In this section we shall be primarily concerned with the mechanisms by which an animal possessing a certain behavior pattern affects the development of the same behavior in a conspecific.

These mechanisms are best studied in animals between whom strong social bonds exist, for in these an individual's behavior may be most readily modified by the behavior of his conspecifics. New responses may appear and old ones may be inhibited or changed. Much of the evidence for such influences stems from observations, such as those of Miyadi (1959) and his co-workers, who observed how first the playmates, then the elders of a particular monkey imitated an accidentally acquired habit of washing sweet potatoes.

There are, of course, many degrees of social interaction in learning situations. Thorpe (1963a) has catalogued these, distinguishing between *social facilitation*, "contagious behavior" (in which a performance by one animal stimulates others to act similarly), *local enhancement* (in which one animal's attention is attracted to a particular locus by the activities of another), and then various kinds of *observational learning* (which may include several distinct kinds of learning processes). It is this last category which is of greatest interest.

Social feeding and observational learning in birds were studied by Turner (1964). He confirmed that birds will copy the feeding habits of others of their kind, eating food or searching in microhabitats that in other circumstances would be unacceptable. Juveniles proved more susceptible to such influences than adults. Interspecific facilitation of feeding could also occur, although intraspecific stimulation was stronger. In another experiment (Klopfer, 1959b, 1961), a discrimination between two sources of food was more readily learned by solitary birds than by pairs or individual birds which had previously observed the training of conspecifics. Even well-trained birds failed to discriminate after having observed the errors of naïve companions. Evidently, the sight of a conspecific feeding, even upon "forbidden fruit" was too potent a signal to be ignored. This is entirely consistent with Turner's findings. In a different species, paired birds did not interfere with one another's learning. It is likely that there are systematic differences with regard to the effects of observation of experienced and naïve conspecifics. It may be that birds with catholic tastes do not interfere or even benefit from one another's experiences, while more conservative feeders prove more immune to social contagions. An alternate explanation was offered by Zajonc (1965) who argued that it was important to distinguish between well-established, dominant responses and newly learned responses. The presence of another individual may enhance performance in the first case but inhibiting it in the latter. However, since Zajonc tried to exclude instances involving imitation or observational learning from his considerations, it is not possible to test his hypothesis with Turner's and Klopfer's data.

Social effects upon the development of pecking and feeding behavior in young precocial birds have been investigated by Hailman (1961, 1964a) on gull chicks and by Tolman (1964) and Tolman and Wilson (1965) on domestic chicks. The gull chick pecks not only at the bill of its parent but also at the white tip of its sibling's bill. Because the pecking movement of the newly hatched chick is somewhat inaccurate (see above), the chick often misses the object at which the peck is aimed. Chicks do not recognize food visually, but may first strike food by missing the bill tip of an experienced, feeding sibling, and thereby striking food. Food in the beak elicits swallowing, and the chick quickly learns to recognize food visually. This sequence was established both by observations from a hide in the wild and by laboratory experiment. The social behavior continues to be of advantage to the developing gull chicks when the parent begins bringing large pieces of food. One chick may peck into the food, which it cannot tear by itself, but a second chick pecking at its bill also may grasp the food, and both tugging in different directions may effect the ripping off of pieces small enough to swallow.

Tolman found a very similar situation in domestic chicks, even though they lack any special bill marking to elicit pecking. Tolman and Wilson have further investigated the role of hunger in both subject and companion as it affects the magnitude of social facilitation in pecking. In both of these examples, the effect of the companion is quite different from the classical case of social facilitation in hens, where the mere sight of a feeding hen will stimulate a previously satiated bird to begin feeding again. Neither of these examples really fits any of Thorpe's categories (above), although they may come closest to the rubric of local enhancement (and see review on observational learning by Davis, 1973).

Finally, examples illustrating the potency of social influences as well as the multiplicity of pathways involved in the control of behavior are to be found in the literature on the development of sexuality (Klopfer and Bernstein, in press; Hampton, 1965).

> The first individual that I've selected from the cases described by Hampton was apparently a normal female at birth and was raised as a female for the first fourteen years of her life. However, even within a couple of months of her birth, her parents commented on her great forcefulness. Her mother was a rather lady-like, demure individual, and commented that her child ate so fast and wasn't at all like a little girl. "She did everything crash bang, nothing gentle." In games, the child took male roles and scarcely could be forced into dresses. She was, however, a very active hiker and participated in school

football games. But at adolescence, she began to quiet down. She developed at this time a persistent hoarseness which occasioned her examination at Johns Hopkins, an examination which ultimately revealed male chromatin, a tiny penis of clitoral size, a presence of undescended testes, all with the external genitals of a female.

Informed that she was genetically a male, her response was a blasé, "I told you so." Subsequently, the family moved and the child, now emerging fully as a male, was accepted as such by his classmates and went on to have dates and orgasm, though lacking a penis and a scrotum. He became known as a superior student. The assignment had originally been to the female class, and the genitals certainly suggested the validity of this assignment. However, the gonads and the chromatin and the child's own psychological state were male. The dissonance was in the biological markers and not in the child's own mental state.

A rather different case was presented by a twenty-five year old male transexual described by Hampton. This individual was seen by him as a dyed blonde with all the mannerisms of a woman. He reported that his earliest recollections were of feeling feminine. He dated men, though he didn't permit genital intimacy, for this apparently led him to identify the partner as a homosexual. This man took estrogens for breast development, underwent electrolysis for the removal of hair, and thus succeeded in holding a number of jobs normally assigned women, secretarial positions which he filled successfully. Finally, he underwent an operation for the removal of his penis and testes and had his scrotal skin invaginated for a vagina. The transformation complete, he—now she—married, engaged in vaginal intercourse with her husband, claimed to experience orgasms, and was quite happy in the thought that she would adopt children even though she could not carry them herself.

Note again that the patient does not hallucinate and knows full well what was involved in transforming the former "him" to a "her." In this instance, the transformation was made and successfully accomplished in the face of biological indicators which all cried "male."

What then do we mean by male and female? It is very likely that what we think we mean has nothing to do with the biological reality. The biological reality may be that there are not merely two sexes but two extremes with regard to sexual status, with individuals forming a continuum from one extreme to the other. This is not a simple continuum but a continuum along a number of different dimensions so that what we are dealing with are varieties of sex. Each individual may be invested with a variety of degrees of maleness or femaleness. The factors that determine where an individual will fall in any one of the continua that represents the various kinds of sexuality that exist may be rather variable. In one case, they may be the kinds of chemical conditions impinging on the developing embryo, in another, on the kinds of social conditions that impinge on the postpartum infant. Some of the influences may have irrevocable effects. For instance, in the case of guinea pigs, when testosterone is administered prenatally, the effect on the brains of the young unborn females is apparently to render them subsequent insensitive to estrogen. Such endrogenized guinea pigs later in life cannot be made to show the cyclic estrus behavior characteristic of females even when injected with large quantities of estrogenic hormones. Thus, a large dose of a prenatal hormone has an organizing effect on the central nervous system which may cause the overt behavior of the animal to be permanently male-like, even while the gonads and the external genitals are unchanged and in the female state.

In the normal course of development there may well be a certain degree of concordance between the different kinds of sexuality so the same degree of maleness is established with regard to hormones and gonads and external morphology. Should there be any kind of deviation in any of the conditions normally impinging on the developing embryo, there may then be a dissynchrony established in the developing system such that the various characteristics no longer develop to the same degree. The extent to which the resulting dissonance may be subsequently corrected or not may of course be expected to vary both with the species and with the nature and degree of the perturbation (Klopfer and Bernstein, in press).

Father Mulligan has written:

There is a wide or narrow range of possible expression in different behaviors, and there is crucial dependence on an expected environment as neural maturation and the behavioral system unfolds. The latter is indeed a system, an adaptive whole which is currently being subjected to continually greater stress by subtraction of components of its expected environment, and the addition of irritants. We can survive perhaps, but can we be healthy in the absence of open space, green things, and the gentle, rhythmic sounds of nature? Do suburbs grow in response to the need for the open space experienced in another period of human history? Are drug use and violence related to overall stress on the human adaptive system, which has passed its tolerance limits, or do these result from peculiar learning conditions today? Does the moral view of human sexuality as essentially procreative make sense considering the contrasts in social structure, physiology, and sexual behavior in man, primates, and mammals? (unpublished manuscript).

The lemurs and gulls we study are lovely beasts. We personally require no further excuse to study them. But, Mulligan's queries are germane to the issues which behavior studies raise. Whether wished for or not, explanatory principles derived from behavior studies are daily advanced to justify or excuse human foibles and behavior. At the very least, we ought to see that such principles are not derived from a faulty conception of behavioral mechanisms, their evolution, and development.

Selected Reading

Hailman, J. P. 1967. The ontogeny of an instinct: The pecking response in chicks of the Laughing Gull (*Larus atricilla* L.) and related species. *Behaviour supplement.*

Postscript

We have tried to show the birth and struggle for life of a new science through the ideas of its major participants. The final sections of chapters in Parts II and III and the chapters of Part IV outline only some of the current ideas and methods that we expect will develop at an accelerating rate. Many other new facts and concepts from studies of evolution, ecology, psychology, and physiology are relevant to the current status and future of ethology. We hope the story presented here has whetted the appetite of ethology's future participants—those who will do the original experiments and who will bring about a new synthesis of behavioral theory.

RECENT ARTICLES
IN SCIENTIFIC AMERICAN
DEALING WITH
ANIMAL BEHAVIOR

It is often useful for the undergraduate student to have available a reading list of relatively nontechnical, yet authoritative, articles to gain a firmer hold on his introduction to a subject. Therefore, in addition to the critical references provided at the end of each chapter, there follows a selected list of recent (1950–1972) articles published in *Scientific American*, arranged according to the "causes and origins" of behavior discussed in Part IV.

General Behavioral Description and Behavioral Maintenance

L. J. Milne and M. J. Milne, Animal courtship, July 1950
N. Tinbergen, Curious behavior of the stickleback, Dec. 1952
D. B. Steinmen, Courtship of animals, Nov. 1954
A. M. Guhl, Social order of chickens, Feb. 1956
A. J. Marshall, Bowerbirds, June 1956
W. H. Thorpe, Language of birds, Oct. 1956
N. Tinbergen, Defense by color, Oct. 1957
J. A. King, Social behavior of prairie dogs, Oct. 1959
S. L. Washburn and I. DeVore, Social life of baboons, June 1961
I. Eibl-Eibesfeldt, Fighting behavior of animals, Dec. 1961
J. B. Calhoun, Population density and social pathology, Feb. 1962
K. von Frisch, Dialects of the bees, Aug. 1962
A. M. Wenner, Sound communication in honeybees, April 1964
J. E. Emlen and R. L. Penney, The Navigation of penguins, Oct. 1966
D. Singh, Urban monkeys, July 1969
H. C. Bennet-Clark and A. W. Ewing, The love song of a fruit fly, July 1970
J. M. Todd, The chemical languages of fishes, May 1971
E. O. Wilson, Animal communication, Sept. 1972
A. J. Premack and D. Premack, Teaching language to an ape, Oct. 1972
H. R. Topoff, The social behavior of army ants, Nov. 1972

Behavioral History

N. L. Munn, The evolution of mind, June 1957
K. Z. Lorenz, The evolution of behavior, Dec. 1958

N. Tinbergen, Evolution of behavior in gulls, Dec. 1960
E. T. Gilliard, The evolution of bowerbirds, Aug. 1963
V. C. Wynne-Edwards, Population control in animals, Aug. 1964
M. E. Bitterman, The evolution of intelligence, Jan. 1965
R. J. Andrew, The origins of facial expressions, Oct. 1965
H. Esch, The evolution of bee language, April 1967
A. Seilacher, Fossil behavior, August 1967

Behavioral Ontogeny

C. J. Warden, Animal intelligence, June 1951
B. F. Skinner, How to teach animals, Dec. 1951
R. A. Butler, Curiosity in monkeys, Feb. 1954
N. Pastore, Learning in the canary, June 1955
E. H. Hess, Imprinting in animals, March 1958
H. F. Harlow, Love in infant monkeys, June 1959
E. J. Gibson and R. D. Walk, The "visual cliff," April 1960
W. C. Dilger, Behavior of lovebirds, January 1962
H. F. Harlow and M. K. Harlow, Social deprivation in monkeys, Nov. 1962
J. B. Best, Protopsychology, Feb. 1963
V. H. Denenberg, Experience and emotional development, June 1963
S. C. Wecker, Habitat selection, October 1964
J. P. Hailman, How an instinct is learned, Dec. 1969
M. R. Rosenzweig, E. L. Bennett, and M. L. Diamond, Brain changes in response to experience, Feb. 1972
N. Geschwind, Language and the brain, April 1972
E. H. Hess, "Imprinting" in a natural laboratory, Aug. 1972
J. S. Rosenblatt, Learning in newborn kittens, Dec. 1972

Behavioral Control

D. R. Griffin, Navigation of bats, Aug. 1950
H. Kalmus, Sun navigation of animals, Oct. 1954
R. W. Sperry, Eye and the brain, May 1956
E. H. Hess, Space perception of the chick, July 1956
N. Guttman, and H. I. Kalish, Experiments in discrimination, Jan. 1958
D. R. Griffin, More about bat "radar," July 1958
H. Grundfest, Electric fishes, Oct. 1960

D. S. Blough, Animal psychophysics, July 1961

E. von Holst and U. St. Paul, Electrically controlled behavior, March 1962

H. N. Lissman, Electric location by fishes, March 1963

E. O. Wilson, Pheromones, May 1963

D. Kennedy, Inhibition in visual systems, July 1963

D. H. Hubel, The visual cortex of the brain, Nov. 1963

W. R. A. Muntz, Vision in frogs, March 1964

D. S. Lehrman, The reproductive behavior of ring doves, Nov. 1964

E. F. MacNichol, Jr., Three-pigment color vision, Dec. 1964

B. B. Boycott, Learning in the octopus, March 1965

K. D. Roeder, Moths and Ultrasound, April 1965

R. Held, Plasticity in sensory-motor systems, Nov. 1965

L. R. Peterson, Short-term memory, July 1966

B. W. Agronoff, Memory and protein synthesis, June 1967

D. M. Wilson, The flight control system of the locust, May 1968

K. Pribram, The neurophysiology of remembering, Jan. 1969

C. Greenwalt, How birds sing, Nov. 1969

A. R. Lucia, The functional organization of the brain, March 1970

E. R. Kandel, Nerve cells and behavior, July 1970

S. Levine, Stress and behavior, Jan. 1971

R. C. Atkinson and R. M. Shiffrin, The control of short-term memory, Aug. 1971

Appendix B

JOURNALS IN WHICH STUDIES OF ANIMAL BEHAVIOR ARE FREQUENTLY REPORTED

It is often helpful for an advanced student of a subject to have a list of pertinent journals in his field, because abstracting periodicals—as good as they may be—are never complete. We have provided the following list of journals, arranged according to categories of their specific purposes. Our criteria for inclusion were that the journal must have contained at least two major papers (excluding notes and letters) on animal behavior during 1964 and 1965. Journals no longer published, but so useful for older references that they cannot be omitted, are marked with an asterisk (*).

The Principal Journals of Animal Behavior

Advances in Animal Behavior (annual, Academic Press)
Animal Behaviour (contination of *British Journal of Animal Behaviour),
    including supplements separately numbered
Behaviour, including supplements separately numbered
Behavioral Science
*Comparative Psychology Monographs
*Journal of Animal Behaviour
Journal of Comparative and Physiological Psychology (continuation of
    *Journal of Comparative Psychology and others).
Perspectives in Ethology (annual; Plenum Press)
Zeitschrift für Tierpsychologie

Journals of General Science

Experientia
Nature
Naturwissenschaften
Proceedings of the National Academy of Sciences (U.S.)
Science

Journals of Biology and Zoology

Acta Biologica Academiae Scientiarum Hungaricae
Acta Zoologica Budapest

Acta Zoologica Cacoviensia (Polska Academia nauk.)
Acta Zoologica Fennica
Akademiia nauk SSSR. Doklady Biol. Sci.
American Midland Naturalist
American Museum Novitates
American Naturalist
Annals of the New York Academy of Sciences
Annals for the Society of Zoology and Botany
     Fennica Vanamo.
Archives Neerlandaise de Zoologie
Biological Bulletin
Biological Reviews
Biologisches Zentralblatt
Bulletin of the American Museum of Natural History
Bulletin of the U.S. National Museum
Canadian Field Naturalist
Canadian Journal of Zoology
Ecological Monographs
Ecology
Ergebnisse der Biologie
Evolution
Folia Biologica
Genetics
Heredity
Japanese Journal of Zoology
Journal of Animal Ecology
Journal of Bombay Natural History Society
Journal of Experimental Biology
Journal of Experimental Zoology
Journal of Genetics
Journal of Theoretical Biology
Journal of Wildlife Management
Proceedings of the Royal Society
Proceedings of the Society of Experimental Biology and Medicine
Proceedings of the Zoological Society of London
     (post volume 145, *see* Journal of Zoology)
Pubblicazione Stazione Zoologica Napoli
Quarterly Review of Biology
Revue Suisse de Zoologie
Systematic Zoology
Transactions of the Linnean Society of New York
University of California Publications in Zoology
Verhandlungen der Deutschen Zoologische Gesellschaft

Zeitschrift für Morphologie und Ökologie der Tiere
Zoologica
Zoologischer Anzeiger
Zoologische Jährbücher. Abt. für allgemeine Zoologie und Physiologie der
    Tiere
Zoology

Journals of Psychology

Acta Psychologica
American Journal of Psychology
American Psychologist
Annual Review of Psychology
British Journal of Psychology
Canadian Journal of Psychology
Journal of the Experimental Analysis of Behavior
Journal of Experimental Psychology
Journal of Genetical Psychology
Perceptual and Motor Skills
Psychological Bulletin
Psychologische Forschung
Psychological Reports
Psychological Review
Quarterly Journal of Experimental Psychology
Zeitschrift für Psychologie

Journals of Physiology and Psychophysics

American Journal of Physiology
Annual Review of Physiology
Brain
Endocrinology
Journal of the Acoustical Society of America
Journal of Comparative Neurology
Journal of General Physiology
Journal of Neurophysiology
Journal of the Optical Society of America
Journal of Physiology

Physiological Reviews
Physiological Zoology
Vision Research
Zeitschrift für allgemeine Physiologie
Zeitschrift für vergleichende Physiologie

Journals Devoted to
Specific Groups of Animals

Alauda
Annals of the Entomological Society of America
Ardea
Auk
Avicultural Magazine
Beiträge zur Vogelkunde
Bird-Banding
Bird Study
British Birds
Condor
Copeia
Dansk Ornithologisk forening Tidsskrift
Emu
Folia Primatologica
Herpetologica
Human Biology
Ibis
Insectes Sociaux
Journal of the Fisheries Research Board of Canada
Journal of Mammalogy
Journal für Ornithologie
Mammalia
Ornis Fennica
Ornithologische Berichte
Oryx
Ostrich
Pacific Coast Avifauna
Die Vogelwarte
Vogelzug
Wilson Bulletin
Zeitschrift für Säugetierkunde

Appendix C | REFERENCES CITED

Adams, D. K. 1931. A restatement of the problem of learning. Brit. J. of Psychol. **22**:150–178.

Adams, D. K. 1962. Foreword *in* P. H. Klopfer [ed.], Behavioral aspects of ecology. Prentice-Hall, Inc., Englewood Cliffs, N.J.

Adey, W. R., J. P. Segund, and R. B. Livingstone. 1957. Corticifugal influences on intrinsic brain stem conduction in cat and monkey. J. Neurophysiol. **20**:1–16.

Allee, W. C. 1938. Cooperation among animals. H. Schuman, New York.

Allee, W. C. 1942. Social dominance and subordination among vertebrates. Biol. Symp. **B**:139–162.

Allee, W. C., A. E. Emerson, O. Park, T. Park, and K. P. Schmidt. 1949. Principles of animal ecology. W. B. Saunders Co., Philadelphia.

Allport, F. H. 1955. Theories of perception and the concept of structure. John Wiley & Sons, Inc., New York.

Altmann, M. 1952. Social behavior of elk, *Cervus canadensis* Nelsoni in the Jackson Hole Area of Wyoming. Behaviour **4**:116–143.

Altmann, M. 1960. The role of juvenile elk and moose in the social dynamics of their species. Zoologica **45**:35–39.

Altmann, S. A. 1965. Sociobiology of rhesus monkeys. II: Stochastics of social communication. J. Theoret. Biol. 8:490–522.

Alverdes, F. 1927. Social life in the animal world. Harcourt, Brace, New York.

Alverdes, F. 1935. The behavior of mammalian herds and packs, pp. 185–206. *In* C. Murchison [ed.], A handbook of social psychology. Clark University Press, Worcester, Mass.

Ambrose, A. [ed.], 1969. Stimulation in early infancy. Academic Press, Inc., New York.

Andrew, R. J. 1956. Normal and irrelevant toilet behaviour in *Emberiza* spp. Brit. J. Anim. Behav. 4:85–91.

Andrew, R. J. 1961. The motivational organization controlling the mobbing calls of the blackbird. Behaviour 18:161–176.

Andrew, R. J. 1963. Effect of testosterone on the behavior of the domestic chick. J. Comp. Physiol. Psychol. 56:933–940.

Andrew, R. J. 1964a. The development of adult responses from responses given during imprinting by the domestic chick. Anim. Behav. 12:542–548.

Andrew, R. J. 1964b. Vocalization in chicks and the concept of "stimulus contrast." Anim. Behav. 12:64–76.

Annals of the New York Academy of Sciences. 1962. Rhythmic functions in the living system. 98:753–1326.

Ardrey, R. 1970. The social contract. Atheneum Publishers, New York.

Ariëns Kappers, C. U., G. C. Huber, and E. C. Crosby. 1936. The comparative anatomy of the nervous system of vertebrates including man. The Macmillan Company, New York.

Armstrong, E. A. 1947. Bird display and behaviour. Oxford University Press, Inc., New York.

Armstrong, E. A. 1950. The nature and function of displacement activities, pp. 361–384. *In* Society for Experimental Biology, Symposia IV, Physiological mechanisms in animal behavior. Academic Press, Inc., New York.

Aronson, E., and S. Rosenbloom. 1971. Space perception in early infancy: Perception within a common auditory-visual space. Science 172:1161–1163.

Aschoff, J. 1962. Tierische Periodik unter dem Einfluss von Zeitgebern. Z. Tierpsychol. 15:1–30.

Babich, F. R., A. L. Jacobsen, S. Bubach, and S. and A. Jacobsen. 1965. Transfer of a response to naïve rats by injection of RNA extracted from trained rats. Science 149:656–657.

Barlow, H. B. 1961. The coding of sensory messages, pp. 331–360. *In* W. H. Thorpe and O. L. Zangwill [eds.], Current problems in animal behavior. Cambridge University Press, New York.

Barlow, J. 1964. Inertial navigation as a basis for animal navigation. J. of Theoret. Biol. 6:76–117.

Bartholomew, C. A., and J .B. Birdsell. 1953. Ecology and the proto-hominids. Amer. Anthropol. **55**:481–498.

Bartholomew, G. D., and N. E. Collias. 1962. The role of vocalizations in the social behavior of the northern elephant seal. Anim. Behav. **10**:7–14.

Barzun, J. 1941. Darwin, Marx, Wagner; critique of a heritage. Little, Brown & Co., Boston.

Bastock, M. 1956. A gene mutation that changes a behaviour pattern. Evolution **10**:421–439.

Bastock, M., D. Morris, and M. Moynihan. 1953. Some comments on conflict and thwarting in animals. Behaviour **6**:66–84.

Bateson, G. 1963. The role of somatic change in evolution. Evolution **17**:529–539.

Bateson, M. C. 1972. Our own metaphor. Alfred A. Knopf, Inc., New York.

Bateson, P. P. G. 1966. The characteristics and context of imprinting. Biol. Rev. **41**:177–220.

Bateson, P. P. G. 1973. Internal influences on early learning in birds. *In* R. A. Hinde and J. S. Hinde [eds.], Constraints on learning. Academic Press, Inc., New York. (in press)

Beach, F. A. 1947. A review of physiological and psychological studies of sexual behavior in mammals. Physiol. Rev. **27**:240–307.

Beach, F. A. 1948. Hormones and behavior. P. B. Hoeber, New York.

Beach, F. A. 1951. Instinctive behavior: Reproductive activities, pp. 387–434. *In* S. S. Stevens [ed.], Handbook of experimental psychology. John Wiley & Sons, Inc., New York.

Beach, F. A. 1955. The descent of instinct. Psychol. Rev. **62**:401–410.

Beach, F. A. [ed.]. 1965. Sex and behavior. John Wiley & Sons, Inc., New York.

Beach, F. A. 1971. Hormonal factors controlling the differentiation, development and display of copulatory behavior in the ramsteigig and related species, pp. 249–296. *In* Biopsychology of development. Academic Press, Inc., New York.

Beach, F. A., D. O. Hebb, G. T. Morgan, and H. W. Nissen. 1960. The neuropsychology of Lashley. McGraw-Hill Book Company, New York.

Beer, C. G. 1961. Incubation and nest building behaviour of black-headed gulls. I: Incubation behaviour in the incubation period. Behaviour **18**:62–106.

Beer, C. G. 1962a. The egg-rolling of black-headed gulls *Larus ridibundus*. Ibis **104**: 388–398.

Beer, C. G. 1962b. Incubation and nest building behaviour of black-headed gulls. II: Incubation behaviour in the laying period. Behaviour **19**:283–304.

Beer, C. G. 1963a. Incubation and nest building behaviour of black-headed gulls. III: The pre-laying period. Behaviour **21**:13–77.

Beer, C. G. 1963b. Incubation and nest building behaviour of black-headed gulls. IV: Nest-building in the laying and incubation periods. Behaviour **21**:155–176.

Beer, C. G. 1963–1964. Ethology—the zoologists approach to behavior. Tuatara, 11:170–177; 12:16–39.

Beer, C. G. 1969. Laughing gull chicks: Recognition of their parents' voices. Science 166:1030–1032.

Beer, G. R. de. 1940. Embryos and ancestors. Clarendon Press, Oxford.

Benzer, S. 1971. From the gene to behavior. J. Amer. Med. Assoc. 218:1015–1022.

Berg, I. A. 1944. Development of behavior: The micturition pattern in the dog. J. Exp. Psychol. 34:343–368.

Bermant, G. 1973. Perspectives in animal behavior. Scott, Foresman & Company, Glenview, Illinois.

Bernhard, C. G., et al. 1970. Eye ultrastructure, colour reception and behaviour. Nature 226:865–866.

Bissonette, T. H. 1933. Light and sexual cycles in starlings and ferrets. Quart. Rev. Biol. 8:201–208.

Bissonette, T. H. 1943. Some recent studies on photoperiodicity in animals. Trans. N.Y. Acad. Sci. 5:43–51.

Black-Cleworth, P. 1970. The role of electric discharges in the non-reproductive social behavior of *Gymnotus cerapo*. Anim. Behav. Monogr. 3:1–77.

Blair, W. F. 1958. Mating call in the speciation of anuran amphibians. Amer. Nat. 92:27–51.

Blodgett, H. C. 1929. The effect of the introduction of reward upon the maze performance of rats. Univ. Calif. Pub. Psychol. 4:113–134.

Boakes, R. A., and M. S. Halliday. 1970. *In* R. Borger and F. Cioffi [eds.], Explanation in the behavioral sciences. Cambridge University, New York.

Bogert, C. M. 1960. The influence of sound on the behavior of amphibians and reptiles. Animal Sounds and Communications, Pub. No. 7:139–316. Amer. Inst. Biol. Sci.

Bonner, J. T. 1955. Cells and societies. Princeton University Press, Princeton, N.J.

Borgreve, B. 1884. *See* Stresemann, E. Die Entwicklung der Ornithologie. F. W. Peters, Berlin.

Boring, E. G. 1957. A history of experimental psychology, 2nd ed. Appleton-Century-Crofts, Inc., New York.

Bossert, W. H., and E. O. Wilson. 1963. The analysis of olfactory communication among animals. J. Theoret. Biol. 5:443–469.

Boycott, B. B. 1954. Learning in *Octopus vulgaris* and other cephalopods. Sta. Zool. Pub. Napoli 25:67–93.

Brady, J. V., and W. J. H. Nauta. 1955. Subcortical mechanisms in emotional behavior. J. Comp. Physiol. Psychol. 48:412–420.

Braemer, W., and H. O. Schwassmann. 1963. Vom Rhythmus der Sonnenorientierung am Aquator (bei Fischen). Ergeb. Biol. 26:183–201.

Broca, Paul. 1861. Remarques sur le siège de la faculté du language articulé, suivi d'une observation d'aphémie (part de la parole). Bull. Soc. Anat. 2 ser. 6:330–357.

Brown, F. A. 1957. The rhythmic nature of life, pp. 287–304. *In* Recent advances in invertebrate physiology. University of Oregon Pub., Eugene.

Brown, F. A. 1959. Living clocks. Science 130:1535–1544.

Brown, J. L. 1964. The integration of agonistic behavior in the Steller's jay *Cyanocitta stelleri* (Gmelin). Univ. Calif. Pub. Zool. 60:223–328.

Buddenbrock, W. von. 1915. Die Tropismentheorie von Jacques Loeb, ein Versuch iher Widerlegung. Biol. Zentralbl. 35:481–506.

Bullock, T. H. 1958. Evolution of neurophysiological mechanisms, pp. 165–178. *In* A. Roe and G. G. Simpson [eds.], Behavior and evolution. Yale University Press, New Haven, Conn.

Bullock, T. H., and R. B. Cowles. 1952. Physiology of an infra-red receptor: the facial pit of pit-vipers. Science 115:541–543.

Bullough, W. S. 1945. Endocrinological aspects of bird behavior. Biol. Rev. 20:89–99.

Bünning, E. 1963. Die Physiologische Uhr. Springer-Verlag, Berlin.

Buttel-Reepen, H. von. 1900. Sind die Beinen "Reflexmaschinen"? Biol. Zentralbl. 20:97–109.

Calhoun, J. B. 1962. A "behavioral sink," pp. 295–315. *In* E. L. Bliss [ed.], Roots of behavior. Harper & Row, Publishers, New York.

Calhoun, J. B. 1963. The social use of space. *In* W. V. Mayer and R. G. van Gelder [eds.], Physiological mammology, Vol. 1. Academic Press, Inc., New York.

Candolle, A. P. de. 1832. Physiologie végétale. Bechet Jeune, Paris.

Cannon, W. B. 1915. Bodily changes in pain, hunger, fear, and rage. W. W. Norton & Company, Inc., New York.

Capranica, R. R. 1965. The evoked vocal response of the bullfrog—a study of communication by sound. Unpublished Sc.D. dissertation, M.I.T.

Carmichael, L. 1926. The development of behavior in vertebrates experimentally removed from the influence of stimulation. Psychol. Rev. 33:51–58.

Carmichael, L. 1927. A further study of development of behavior in vertebrates experimentally removed from the influence of external stimulation. Psychol. Rev. 34:34–47.

Carmichael, L. 1934. An experimental study in the prenatal guinea pig of the origin and development of reflexes and patterns of behavior in relation to the stimulation of specific receptor areas during the period of active fetal life. Genet. Psychol. Monogr. 16:338–491.

Carpenter, C. R. 1934. A field study of the behavior and social relations of howling monkeys. Comp. Psychol. Monogr. 10:1–168.

Carpenter, C. R. 1940. A field study of the behavior and social relations of the gibbon. Comp. Psychol. Monogr. 16:1–212.

Carthy, J. D. 1956. Animal navigation: How animals find their way about. George Allen & Unwin, London.

Cheal, M. L., and R. L. Sprott. 1971. Social olfaction: A review of the role of olfaction in a variety of animal behaviors. Psychol. Reports, Monogr. Suppl. 1–V29, 29:195–243.

Chepko, B. D. 1971. A preliminary study of the effects of play deprivation on young goats. Zeit. f. Tierpsych. 28:517–526.

Chesler, P. 1969. Maternal influence in learning by observation in kittens. Science 166:901–903.

Christian, J. J. 1963. Endocrine adaptive mechanisms and the physiologic regulation of population growth, pp. 189–353. In W. V. Mayer and R. G. van Gelder [eds.], Physiological mammalogy. Academic Press, Inc., New York.

Christian, J. J. 1970. Social subordination, population density, and mammalian evolution. Science 168:84–90.

Cloudsley-Thompson, I. L. 1961. Rhythmic activity in animal physiology and behaviour. Academic Press, Inc., New York.

Coghill, G. E. 1929. Anatomy and the problem of behaviour. Cambridge University Press, New York.

Cohen, J. E. 1969. Natural primate troops and a stochastic population model. Amer. Nat. 103:455–479.

Cold Spring Harbor Symposium. On quantitative biology. 1960. Vol. 25.

Cole, L. C. 1957. Biological clock in the unicorn. Science 125:874–876.

Collias, N. E. 1944. Aggressive behavior among vertebrate animals. Physiol. Zool. 17:83–123.

Collias, N. E. 1951. Social life and the individual among vertebrate animals. Ann. N.Y. Acad. Sci. 51:1074–1092.

Collias, N. E. 1956. The analysis of socialization in sheep and goats. Ecology 37:228–239.

Collins, T. B. 1965. Strength of the following-response in the chick in relation to degree of "parent" contact. J. Comp. Physiol. Psychol. 60:192–195.

Corning, W. C., and E. R. John. 1961. Effect of ribonuclease on retention of conditioned response in regenerated planarians. Science 134:1363–1365.

Correns, C. 1907. Die Bestimung und Vererbung des Geschlechts nach neuen Versuchen mit hoheren Planzen. Gebrüder Borntraeger, Berlin.

Craig, W. 1918. Appetites and aversions as constituents of instincts. Biol. Bull. 34:91–107.

Crew, F. A. E. 1923. Studies of intersexuality. II: Sex reversal in the fowl. Roy. Soc. (London). Proc. 95:265–278.

Crew, F. A. E. 1953. Sex determination. Methuen & Co., Ltd., London.

Crook, J. H. 1970. Social behaviour in birds and mammals. Academic Press, Inc., New York.

Crozier, W. J., and H. Hoagland. 1934. The study of living organisms, pp. 3–108. *In* C. Murchison [ed.], Handbook of general experimental psychology. Clark University Press, Worcester, Mass.

Cruze, W. W. 1935. Maturation and learning in chicks. J. Comp. Psychol. **19**:371–409.

Cullen, E. 1957. Adaptations in the kittiwake to cliff-nesting. Ibis **99**:275–302.

Curio, E. 1961. Zur geographischen Variation von Verhaltensweisen. Die Vogelwelt **2**:33–47.

Daanje, A. 1950. On locomotory movements in birds and the intention movements derived from them. Behaviour **3**:48–99.

Dane, B., and W. G. van der Kloot. 1964. An analysis of the display of the goldeneye duck, *Budephala clangula*. Behaviour **22**:282–328.

Dane, B., C. Walcott, and W. H. Drury. 1959. The form and duration of the display actions of the goldeneye. Behaviour **14**:265–281.

Darling, F. F. 1937. A herd of red deer. Cambridge University Press, London.

Darling, F. F. 1938. Bird flocks and the breeding cycle. Cambridge University Press, London.

Darwin, C. 1859. The origin of species. D. Appleton & Co., New York.

Darwin, C. 1868. Variation of animal and plants under domestication. D. Appleton & Co., New York.

Darwin, C. 1871. The descent of man. D. Appleton & Co., New York.

Darwin, C. 1873. Expression of the emotions in man and animals. D. Appleton & Co., New York.

Darwin, C. 1876. The movements and habits of climbing plants. D. Appleton & Co., New York.

Davis, M. 1973. Imitation: a review and critique. *In* P. P. G. Bateson and P. H. Klopfer [eds.], Perspectives in ethology. Plenum Press, New York. (in press)

Deegener, P. 1918. Die Formen der Vergesellschaftung im Tierreiche. Veit, Leipzig.

Denenberg, V. H. 1964. Critical periods, stimulus input and emotional reactivity: A theory of infantile stimulation. Psychol. Rev. **71**:335–351.

Denenberg, V. H., G. A. Hudgens, and M. X. Zarrow. 1964. Mice reared with rats: Modification of behavior by early experience with another species. Science **143**:380–381.

Denenberg, V. H., and A. E. Whimbey. 1963. Behavior of adult rats is modified by the experiences their mothers had as infants. Science **142**:1192–1193.

Dethier, V. G. 1955. The physiology and histology of the contact chemoreceptors of the blowfly. Quart. Rev. Biol. 30:348–371.

Dethier, V. G., and E. Stellar. 1961. Animal behavior: Its evolutionary and neurological basis. Prentice-Hall, Inc., Englewood Cliffs, N.J.

Deutsch, J. A. 1960. The structural basis of behavior. University of Chicago Press, Chicago.

Deutsch, J. A. 1969. The physiological basis of memory. Ann. Rev. Psychol. 20:85–104.

Diamond, M. 1970. Intromission pattern and specific vaginal code in relation to induction of pseudo-pregnancy. Science 169:995–997.

Dice, L. R. 1945. Minimum intensities of illumination under which owls can find dead prey by sight. Amer. Natur. 79:385–416.

Dijkgraf, S. 1946. Die Sinneswelt der Fledermäuse. Experientia 2:438–448.

Dilger, W. C. 1960. The comparative ethology of the African parrot, genus *Agapornis*. Z. Tierpsychol. 17:649–685.

Doncaster, L., and G. H. Raynor. 1906. On breeding experiments with Lepidoptera. Zool. Soc. (London), Proc. 6:125–133.

Donner, K. O. 1953. The spectral sensitivity of the pigeon's retinal elements. J. Physiol. 122:524–537.

Dorst, J. 1962. The migration of birds. Houghton Mifflin Company, Boston.

Drews, D. R. 1973. Group formation in captive *Galago crassicaudatus*: Notes on the dominance concept. Zeitschrfit für Tierpsychol. (in press)

Ebbinghaus, H. 1913. Memory; a contribution to experimental psychology. [Trans. by H. A. Ruger and C. E. Bussenius.] Columbia University Press, New York.

Eccles, J. C. 1953. The neurophysiological basis of mind. Oxford University Press Inc., New York.

Eccles, J. C. 1957. The physiology of nerve cells. Johns Hopkins Press, Baltimore.

Eccles, J. C. 1964. The physiology of synapses. Academic Press, Inc., New York.

Eccles, J. C. [ed.]. 1966. Brain and conscious experience. Springer-Verlag, Heidelberg.

Eiduson, S., E. Gelles, A. Yuwiller, and B. Eiduson. 1964. Biochemistry and behavior. D. Van Nostrand Co., Inc., Princeton, N.J.

Eisenberg, J. F., N. A. Muckenhirn, and R. Rudran. 1972. The relation between ecology and social structure in primates. Science 176:863–874.

Ellis, P. 1959. Learning and social aggregation in locust hoppers. Anim. Behav. 7:91–106.

Elton, C. 1942. Voles, mice and lemmings. Oxford University Press Inc., New York.

Emerson, A. E. 1938. Termite nests—a study of the phylogeny of behavior. Ecol. Monogr. 8:237–284.

Emlen, J. T., Jr. 1963. Determinants of cliff edge and escape responses in herring gull chicks in nature. Behaviour 22:1–15.

Emlen, J. T., Jr., and F. W. Lorenz. 1942. Pairing responses of free-living valley quail to sex hormone implants. Auk 59:369–378.

Emlen, S. 1969. Bird migration: Influence of physiological state upon celestial orientation. Science 165:716–718.

Erickson, C. J., and D. S. Lehrman. 1964. Effect of castration of male ring doves upon ovarian activity of females. J. Comp. Physiol. Psychol. 58:164–166.

Erickson, R. P. 1963. Sensory neural patterns and gustation, pp. 205–213. *In* Int. Symp. Olfaction and Taste, Proc. I. Pergamon Press, New York.

Espinas, A. 1878. Des societés animales. Baillière, Paris.

Esser, A. H. 1971. Behavior and environment. Plenum Press, New York.

Ewing, A. W. 1961. Body size and courtship behaviour in *Drosophila melanogaster*. Anim. Behav. 9:93–99.

Exner, S. 1893. Negativ Versuchsergebnisse ueber das Orientierungsvermoegen der Brieftauben. Sitz. Ber. Akad. Wiss. Wien; math.-naturwiss.-liche Klass., Abt. III, 102:318–331.

Fabre, J. H. 1916. The hunting wasps. Hodder & Stoughton, Ltd., London.

Fabre, J. H. 1920. The wonders of instinct. George Allen & Unwin, London.

Farner, D. S. 1959. Photoperiodic control of annual gonadal cycles in birds, pp. 717–750. *In* American Association for the Advancement of Science, Photoperiodism and related phenomena in plants and animals. Washington, D.C.

Farner, D. S. 1961. Comparative physiology: Photoperiodicity. Ann. Rev. Physiol. 23:71.

Farner, D. S. and J. R. King [eds.]. 1971. Avian biology, pp. 1–556. Academic Press, Inc., New York.

Fentress, J. C. 1973. Interactions between specific and nonspecific factors in the causation of behavior: a conceptual-operational approach. *In* P. P. G. Bateson and P. H. Klopfer [eds.], Perspectives in ethology. Plenum Press, New York. (in press)

Ferrier, D. 1876. The functions of the brain. G. P. Putnam's Sons, New York.

Fischer, K. 1960. Dressur von Smargdeidechsen auf Kompass Richtungen. Die Naturwissenschaften 4:93–94.

Fisher, A. E. 1958. Effects of stimulus variation on sexual satiation in the male rat. Amer. Psychol. 13:382 (Abstr.).

Fisher, J. 1954. Evolution and bird sociality, pp. 71–83. *In* J. A. Huxley, A. S. Hardy, and E. B. Ford [eds.], Evolution as a process. George Allen & Unwin, London.

Fisher, R. A. 1930. The genetical theory of natural selection. Oxford University Press, London.

Flexner, J. B., and L. B. Flexner. 1963. Memory in mice as affected by intracerebral puromycin. Science 141:54–59.

Folley, S. J., and G. S. Knaggs. 1965. Levels of oxytocin in the jugular vein blood of goats during parturition. J. Endocrinol. 33:301–316.

Forel, A. H. 1904. Ants and some other insects. Open Court Publishing Co., Chicago.

Forel, A. H. 1910. Das Sinnesleben der Insekten. E. Reinhardt, Munich.

Foss, B. M. [ed.]. 1965. Determinants of infant behaviour, Vol. 3. Methuen & Co., Ltd., London.

Fox, M. W. 1971. Integrative development of brain and behavior in the dog. University of Chicago Press, Chicago.

Fraenkel, G. S., and D. L. Gunn. 1940. The orientation of animals. Oxford University Press, London.

Freedman, S. 1973. Orientation of birds by geomagnetic field. In P. P. G. Bateson P. H. Klopfer [eds.], Perspectives in ethology. Plenum Press, New York. (in press)

Friedman, H. 1935. Bird societies, pp. 142–184. In C. Murchison [ed.], A handbook of social psychology. Clark University Press, Worcester, Mass.

Frisch, K. von. 1923. Uber die "Sprache" der Bienen. Zool. Jahreb. 40(3):1–186.

Frisch, K. von. 1927. Aus dem Leben der Bienen. Springer-Verlag, Berlin.

Frisch, K. von 1955. The dancing bees. [Transl. by D. Ilse.] Harcourt, Brace and Jovanovich, Inc., New York.

Frishkopf, L. S., and M. H. Goldstein. 1963. Responses to acoustic stimuli from single units in the eighth nerve of the bullfrog. J. Acoust. Soc. Amer. 35:1219–1228.

Fritsch, G., and E. Hitzig. 1870. Ueber die elektrische Erregbarkeit des Grosshirns. Archives Anatomisches Physiologie, pp. 300–332.

Fromme, H. G. 1961. Untersuchungen über das Orientierungsvermögen nächtlich ziehender Kleinvögel (Erithacus rubecula, Sylvia communis). Z. Tierpsychol. 18:205–220.

Galambos, R. 1959. Electrical correlates of conditioned learning, pp. 375–416. In Josiah Macy Foundation, Trans. of First Conference, The central nervous system and behavior.

Galler, S. R. et al. [eds.]. 1972. Animal orientation and navigation. NASA Wash., D.C.

Galton, F. 1899. Natural inheritance. Macmillan & Co., Ltd., London.

Garcia, J. 1973. Natural responses to scheduled rewards. In P. P. G. Bateson and P. H. Klopfer [eds.], Perspectives in ethology. Plenum Press, New York. (in press)

Gause, G. J. 1942. The relation of adaptability to adaptation. Quart. Rev. Biol. 17:99–114.

Gaze, R. M., and M. J. Keating. 1972. The visual system and "neuronal specificity." Nature, 273:375–378.

Geisler, C., W. A. van Bergeijk, and L. S. Frishkopf. 1964. The inner ear of the bullfrog. J. Morph. 114:43–58.

Gerdes, K. 1962. Richtungstendenzen vom Brutplatz verfrachteter Lachmöven (Larus ridibundus L.) unter Ausschluss visueller Gelände und Himmelsmarken. Z. Wiss. Zool. 166:352–410.

Ghent, A. W. 1960. A study of the group-feeding behavior of larvae of the Jack Pine sawfly, Neodiprion pratti. Behaviour 16:110–148.

Ginsberg, B. E. 1965. Coaction of genetical and nongenetical factors influencing sexual behavior, pp. 53–75. In F. A. Beach [ed.], Sex and behavior. John Wiley & Sons, Inc., New York.

Glassman, E. 1969. The biochemistry of learning: An evaluation of the role of RNA and protein. Ann. Rev. Biochem. 38:605–646.

Goethe, F. 1937. Beobachtungen und untersuchungen zur Biologie der Silbermöwe auf der Vogelinsel Memmertsand. J. Ornithol. 85:1–119.

Golani, I. 1973. Non-metric analysis of behavioral interaction sequences in captive jackals. Behaviour 44:89–112.

Golani, I., and H. Mendelssohn. 1971. Sequences of pre-copulatory behavior of the jackal (Canis aureus). Behaviour 38:169–192.

Goldschmidt, R. 1923. The mechanism and physiology of sex-determination. Methuen & Co., Ltd., London.

Goldsmith, T. H. 1960. The nature of the retinal action potential and the spectral sensitivities of ultraviolet and green receptor systems of the compound eye of the worker honeybee. J. Gen. Physiol. 43:775–799.

Gorski, R. A. 1971. Gonadal hormones and the perinatal development of neuroendocrine function, pp. 237–290. In L. Martini and W. Ganong [eds.], Frontiers in neuroendocrinology. Oxford University Press, New York.

Goss-Custard, J. D. 1970. Factors affecting the diet and feeding rate of the redshank. pp. 101–110. In A. Watson [ed.], Animal populations in relation to their food resources. Blackwell Scientific Pubs., Oxford.

Gottlieb, G. 1961. The following-response and imprinting in wild and domestic ducklings of the same species. Behaviour 18:205–228.

Gottlieb, G. 1965. Imprinting in relation to parental and species identification by avian neonates. J. Comp. Physiol. Psychol. 59:345–356.

Gottlieb, G. 1970a. Conceptions of prenatal behavior, pp. 111–137. In I. R. Aronson, E. Tobach, D. S. Lehrman, and J. S. Rosenblatt [eds.], Development of evolution of behavior. W. H. Freeman and Co., Publishers, San Francisco.

Gottlieb, G. 1970b. Development of species identification in birds: An inquiry into the prenatal determinants of perception. University of Chicago Press, Chicago.

Granit, R. 1942. Colour receptors of the frog's retina. Acta Physiol. Scand. 3:137–151.

Granit, R. 1947. Sensory mechanisms of the retina. Oxford University Press, London.

Granit, R. 1955. Receptors and sensory perception. Yale University Press, New Haven, Conn.

Gray, P. H. 1962. Douglas Alexander Spalding: The first experimental behaviorist. J. Gen. Psychol. 67:299–307.

Griffin, D. R. 1944. The sensory basis of bird migration. Quart. Rev. Biol. 19:21–32.

Griffin, D. R. 1952. Bird navigation. Biol. Rev. 27:359–400.

Griffin, D. R. 1958. Listening in the dark. Yale University Press, New Haven, Conn.

Griffin, D. R. 1959. Echoes of bats and men. Doubleday & Company, Garden City, New York.

Griffin, D. R. 1969. The physiology and geophysics of bird navigation. Quart. Rev. Biol. 44:255–276.

Grinnell, A. D. 1963. The neurophysiology of audition in bats. J. Physiol. 167:38–96.

Grunt, J. A., and W. C. Young. 1952. Psychological modification of fatigue following orgasm (ejaculation) in the male guinea pig. J. Comp. Physiol. 45:508.

Grüsser, O. J., and U. Grüsser-Cornehls. 1968. Neurophysiologische Grundlagen visueller angeborener Auslösemechanismus beim Frosch. Zeit. f. vergl. Physiologie 59:1–24.

Gualitierotti, T., B. Schreiber, D. Mainardi, and D. Passerini. 1959. Effect of acceleration on cerebellar potentials in birds and its relation to sense of direction. Amer. J. Physiol. 197:469–474.

Gunther, M. 1961. Infant behaviour at the breast, pp. 37–44. In B. M. Foss [ed.], Determinants of infant behaviour. Methuen & Co. Ltd., London.

Guthrie, E. R. 1935. The psychology of learning. Harper & Row, Publishers, New York.

Guthrie, R. 1959. Association by contiguity, pp. 158–195. In S. Koch [ed.], Psychology: A study of a science, Vol. 2. McGraw-Hill Book Company, New York.

Guttman, N., and H. I. Kalish, 1956. Discriminability and stimulus generalization. J. Exp. Psychol. 51:79–88.

Guttman, R., I. Lieblich, and G. Naftali. 1969. Variation in activity score and sequences in two inbred mouse strains, their hybrids, and back crosses. Anim. Behav. 17:374–385.

Hachet-Souplet, P. 1909. Quelques expériences nouvelles sur les pigeons voyageurs, pp. 663–667. In 6th Int. Congr. Psychol., Proc.

Hailman, J. P. 1961. Why do gull chicks peck at visually contrasting spots? A suggestion concerning social learning of food discrimination. Amer. Natur. 95:245–247.

Hailman, J. P. 1964a and 1967. Ontogeny of an instinct. Ph.D. Thesis. Duke University, Durham, N.C. And Behaviour Supplement. E. J. Brill, Leiden.

Hailman, J. P. 1964b. Breeding synchrony in the equatorial swallow-tailed gull. Amer. Natur. 98:79–83.

Hailman, J. P. 1964c. Coding of the colour preference of the gull chick. Nature **204**: 710.

Hailman, J. P. 1965. Cliff-nesting adaptations in the Galapagos swallow-tailed gull. Wilson Bull. **77**:346–362.

Hailman, J. P. 1966. Mirror-image color-preferences for background and stimulus-object in gull chicks. Experientia **22**:257.

Haldane, J. B. S. 1954. Introducing Douglas Spalding. Brit. J. Anim. Behav. **2**:1–2.

Hale, E. B., and J. O. Almquist. 1960. Relation of sexual behavior to germ cell output in farm animals. J. Dairy Sci. (suppl.) **43**:145–169.

Hall, C. S. 1951. The genetics of behavior, pp. 304–329. In S. S. Stevens [ed.], Handbook of experimental psychology. John Wiley & Sons, Inc., New York.

Hall, K. R. L. 1963. Variations in the ecology of the Chacma baboon, *Papio ursinus*. Symp. Zool. Soc. London **10**:1–28.

Hall, K. R. L. 1965. Social organization of the old world monkeys and apes. Symp. Zool. Soc. London **14**:265–289.

Hamburger, U. 1968. Emergence of nervous coordination. Origins of integrated behavior. Develop. Biol. Supple. **2**:251–271.

Hamburgh, M. 1971. Theories of differentiation. American Elsevier Pub. Co., Inc., New York.

Hamilton, W. D. 1963. The evolution of altruistic behavior. Amer. Natur. **97**:354–356.

Hamilton, W. D. 1964. The genetical evolution of social behaviour I and II. J. Theoret. Biol. **7**:1–16; 17–52.

Hamilton, W. F., and T. B. Coleman. 1933. Trichromatic vision in the pigeon as illustrated by the spectral discrimination curve. J. Comp. Psychol. **15**:183–191.

Hampton, J. L. 1965. Determinants of psychosexual orientation, pp. 108–132. *In* F. A. Beach [ed.], Sex and Behavior. John Wiley & Sons, Inc., New York.

Hansen, E. W. 1971. Squab induced crop growth in experienced and inexperienced ring dove foster parents. J. Comp. Physiol. Psychol. **77**:375–381.

Harlow, H. F. 1949. The formation of learning sets. Psychol. Rev. **56**:51–65.

Harlow, H. F. 1965. Sexual behavior in the rhesus monkey, pp. 234–265. *In* F. A. Beach [ed.] Sex and behavior. John Wiley & Sons, Inc., New York.

Harlow, H. F., and M. K. Harlow. 1962. The effect of rearing conditions on behavior. Bull. Menninger Clinic **26**:213–224.

Harmon, L. D. 1964. Problems in neural modeling, pp. 9–30. *In* R. F. Reiss [ed.], Neural theory and modeling. Stanford University Press, Calif.

Harris, G. W., R. P. Michael, and P. P. Scott. 1958. Neurological basis of behavior. Ciba Found. Symp. Churchill, London.

Hartline, H. K. 1938. The response of single optic nerve fibers of the vertebrate eye to illumination of the retina. Amer. J. Physiol. **121**:400–415.

Haskins, C. P., and E. F. Haskins, 1958. Note on the inheritance of behavior patterns for food selection and cocoon spinning in F_1 hybrids of *Callosamia promethea* and *C. angulifera*. Behaviour 13:89–95.

Hebb, D. O. 1949. The organization of behavior: A neuropsychological theory. John Wiley & Sons, Inc., New York.

Hebb, D. O. 1961. Distinctive features of learning in the higher animals, pp. 37–57. *In* J. F. Delafresnave [ed.], Brain mechanisms and learning. C.I.O.M.S. Symp., Oxford.

Hediger, H. 1955. Studies of the psychology and behavior of captive animals in zoos and circuses. [Transl. by G. Sircom.] Criterion Books, Inc., New York.

Heinroth, O. 1910. Beitrage zur Biologie, namentlich Ethologie und Physiologie der Anatiden. 5 International Ornithologisches Kongress, Verh. 5:589–702.

Helmholtz, H. L. F. 1852. Ueber die Theorie der zusammengestzen Farben. Wiss. Abhandl. II:3–23.

Helms, K. W., and W. H. Drury, 1960. Winter and migratory weight and fat field studies in some North American buntings. Bird-Banding 31:1–40.

Herre, W. 1955. Fragen und Ergebnisse der Domestikationsforschung nach Studien am Hirn. Deut. Zool. Ges. Verhandl. p. 144–214.

Hess, E. 1964. Imprinting in birds. Science 146:1128–1139.

Hess, W. R. 1949. Das Zwischenhirn: Syndrome, Lokalisationen, Funktionen. Schwate, Basel.

Heusser, H. 1960. Über die Beziehung der Erdkröte zu ihrem Laichplatz. Behaviour 16:93–109.

Hinde, R. A. 1954. Changes in responsiveness to a constant stimulus. Brit. J. Anim. Behav. 2:41–55.

Hinde, R. A. 1956a. The behaviour of certain cardueline F_1 inter-species hybrids. Behaviour 9:202–213.

Hinde, R. A. 1956b. Ethological models and the concept of drive. Brit. J. Phil. Sci. 6:321–331.

Hinde, R. A. 1958. The nest-building behaviour of domesticated canaries. Zool. Soc. (London) Proc. 131:1–48.

Hinde, R. A. 1959a. Unitary drives. Anim. Behav. 7:130–141.

Hinde, R. A. 1959b. Motivation. Ibis 101:353–357.

Hinde, R. A. 1960a. Factors governing the changes in strength of a partially inborn response, as shown by the mobbing behaviour of the chaffinch (*Fringilla coellebs*). Royal Society (London) Proc. B. 153:398–420.

Hinde, R. A. 1960b. Energy models of motivation. Symp. Soc. Exp. Biol. 14:199–213.

Hinde, R. A. 1970. Animal behaviour. McGraw-Hill Book Company, New York.

Hinde, R. A. 1972. Nonverbal communication. Cambridge University Press.

Hinde, R. A., and J. S. Hinde [eds.]. 1973. Constraints on learning. Academic Press, Inc., New York.

Hirsch, J. 1962. Individual differences in behavior and their genetic basis, pp. 3–23. *In* E. L. Bliss [ed.] Roots of behavior. Hoeber, New York.

Hochbaum, H. A. 1955. Travels and traditions of waterfowl. University of Minnesota Press, Minneapolis.

Hodgson, E. S. 1961. Taste receptors. Sci. Amer. **204**:135–144.

Hoebel, B. G., and P. Teitelbaum. 1962. Hypothalamic control of feeding and self-stimulation. Science **135**:375–377.

Hoffman, K. 1953. Experimentelle Aenderung des Richtungsfinden beim Star durch Beeinflussung der "inneren Uhr." Naturwissenschaften **40**:608–609.

Holmes, S. J. 1905. The selection of random movements as a factor in phototaxis. J. Comp. Neurol. **15**:98–112.

Holst, E. von. 1950. Quantitative Messung von Stimmungen im Verhalten der Fische. *In* Physiological mechanisms in animal behavior. Symp. Soc. Exp. Biol. **14**:143–174.

Holst, E. von. 1954. Relations between the central nervous system and the peripheral organs. Brit. J. Anim. Behav. **2**:89–94.

Holst, E. von, and H. Mittelstaedt. 1950. Das Reafferenzprinzip. Naturwissenschaften **20**:464–476.

Holst, E. von, and U. St. Paul. 1960. Vom Wirkungsgefüge der Triebe. Naturwissenschaften **47**:409–422.

Holst, E. von, and U. St. Paul. 1963. On the functional organization of drives. J. Anim. Behav. **11**:1–20.

Honig, W. K., and P. H. R. James [eds.]. 1971. Animal memory. Academic Press, Inc., New York.

Horn, G. 1965. Physiological and psychological aspects of selective perception, pp. 155–216. *In* D. Lehrman et al. [eds.], Advances in the study of behavior, Vol. 1. Academic Press, Inc., New York.

Horridge, G. A. 1968. Interneuerons. W. H. Freeman & Co., Publishers, San Francisco.

Howard, E. 1920. Territory in bird life. John Murray, Publishers, Ltd., London.

Hubel, D. H., and N. Wiesel. 1959. Receptive fields of single neurons in the cat's striate cortex. J. Physiol. **148**:574–591.

Hubel, D. H., and N. Wiesel. 1965. Receptive fields and functional architecture in two non-striate visual areas of the cat. J. Neurophysiol. **28**:229–289.

Hull, C. L. 1943. Principles of behavior. Appleton-Century-Crofts, Inc., New York.

Humphrey, G. 1930. Le Chatelier's rule and the problem of habituation and dehabituation in *Helix albdabris*. Psychol. Forsch. **13**:113–127.

Hutchinson, G. E. 1965. The ecological theatre and the evolutionary play. Yale University Press, New Haven, Conn.

Huxley, J. S. 1914. The courtship habits of the great crested grebe (*Podiceps cristatus*); with an addition to the theory of sexual selection. Zool. Soc. (London) Proc. 2:491–562.

Huxley, J. S. 1941. Genetic interactions in a hybrid pheasant. Zool. Soc. (London) Proc. 111A:41–43.

Iersel, J. J. A. van, and A. C. A. Bol. 1958. Preening of two tern species. A study on displacement activities. Behaviour 13:1–88.

Jacob, F., and J. Monod. 1961. Genetic regulatory mechanisms in the synthesis of proteins. J. Molec. Biol. 3:318–356.

James, W. 1890. Principles of psychology. Henry Holt & Co., New York.

Jaynes, J. 1969. The historical origins of "ethology" and "comparative psychology." Anim. Behav. 17:601–606.

Jennings, H. S. 1906. Behavior of the lower organisms. Columbia University Press; The Macmillan Co., agents, New York.

Jewell, P. A., and C. Loizos, 1966. Play, exploration and territory in mammals. Academic Press, Inc., New York.

John, E. Roy. 1967. Mechanisms of memory. Academic Press, Inc., New York.

Kagan, J., and F. H. Beach. 1953. Effects of early experience on mating behavior in male rats. J. Comp. Physiol. Psychol. 46:204–208.

Kaufman, I. C., and R. A. Hinde. 1961. Factors influencing distress calling in chicks with special reference to temperature changes in social isolation. Anim. Behav. 9:197–204.

Kavanau, J. L. 1963. Compulsory regime and control of environment in animal behavior: I. Wheel-running. Behaviour 20:251–281.

Keeley, K. 1962. Prenatal influence on behavior of offspring of crowded mice. Science 135:44–45.

Keeton, W. T. 1971. Magnets interfere with pigeon homing. Proc. Nat. Acad. Sci. 68:102–106.

Keith, J. B. 1963. Wildlife's ten year cycle. University of Wisconsin Press, Madison.

Kendeigh, S. C., G. C. West, and G. W. Cox. 1960. Annual stimulus for spring migration in birds. Anim. Behav. 8:180–185.

Kettlewell, H. B. D. 1965. Insect survival and selection for pattern. Science 148:1290–1296.

King, J. 1955. Social behavior, social organization, and population dynamics in a black-tailed prairie dog town in the Black Hills of South Dakota. Contributions of the Laboratory of Vertebrate Biology, No. 67. University of Michigan Press, Ann Arbor.

King, J. A. 1958. Parameters relevant to determining the effect of early experience upon the adult behavior of animals. Psychol. Bull. 55:46–58.

Kirkman, V. R. 1937. Bird behaviour. Thomas Nelson & Sons, London and Edinburgh.

Kleitman, N. 1949. Biological rhythms and clocks. Physiol. Rev. 29:1–30.

Klopfer, P. H. 1959a. An analysis of learning in young anatidae. Ecology 40(1):90–102.

Klopfer, P. H. 1959b. Social interactions in discrimination learning with special reference to feeding behaviour in birds. Behaviour 14:282–299.

Klopfer, P. H. 1961. Observational learning in birds. Behaviour 97:71–80.

Klopfer, P. H. 1964. Parameters of imprinting. Amer. Natur. 98:175–182.

Klopfer, P. H. 1965a. Behavioral aspects of habitat selection: A preliminary report on stereotypy in foliage preferences of birds. Wilson Bull. 77:376–381.

Klopfer, P. H. 1965b. Imprinting: A reassessment. Science 147:302–303.

Klopfer, P. H. 1969. Habitats and territories. Basic Books, Inc., New York.

Klopfer, P. H. 1971. Mother love: What turns it on? Amer. Sci. 59:404–407.

Klopfer, P. H. 1973a. Behavioral aspects of ecology, 2nd ed. Prentice-Hall, Inc., Englewood Cliffs, N.J.

Klopfer, P. H. 1973b. Does behavior evolve? Ann. N.Y. Acad. Sci. (in press)

Klopfer, P. H. 1973c. On behavior: Instinct is a Cheshire cat. J. B. Lippincott, Philadelphia.

Klopfer, P. H., D. K. Adams, and M. S. Klopfer. 1964. Maternal "imprinting" in goats. Nat. Acad. Sci. (U.S.) Proc. 52:911–914.

Klopfer, P. H., and M. Bernstein. Biology: A natural esthetics. (in press)

Klopfer, P. H., and J. Gamble. 1967. Maternal imprinting in goats: the role of chemical senses. Z. Tierpsychol. 23:588–592.

Klopfer, P. H., and B. K. Gilbert. 1967. A note on retrieval and recognition of young in the elephant seal. Zeit. f. Tierpsych. 23:757–760.

Klopfer, P. H., and G. Gottlieb. 1962a. Imprinting and behavioral polymorphism: auditory and visual imprinting in domestic ducks and the involvement of the critical period. J. Comp. Physiol. Psychol. 65:126–130.

Klopfer, P. H., and G. Gottlieb. 1962b. Learning ability and behavioral polymorphism within individual clutches of wild ducklings. Z. Tierpsychol. 19:183–190.

Klopfer, P. H., and J. P. Hailman. 1964. Basic parameters of following and imprinting in precocial birds. Z. Tierpsychol. 21:755–762.

Klopfer, P. H., and J. P. Hailman. 1965. Habitat selection in birds, pp. 279–303. *In* D. S. Lehrman, R. A. Hinde, and E. Shaw [eds.], Advances in the study of behavior. Academic Press, Inc., New York.

Klopfer, P. H., and M. S. Klopfer. 1973. How are social roles and ranks determined? Amer. Sci. (in press)

Klüver, H. 1933. Behavior mechanisms in monkeys. University of Chicago Press, Chicago.

Knight, T. A. 1806: On the direction of the radicle and germin vegetation of seeds, Roy. Soc. (London), Phil. Trans. : 99–108.

Knorr, V. A. 1954. The effect of radar on birds. Wilson Bull. 66:264.

Koch, S. [ed.]. 1959. Psychology: A study of a science, Vol. 2: General systematic formulations, learning, and special processes. McGraw-Hill Book Company, New York.

Koehler, O. 1950. Die Analyse der Taxisanteile instinktartigen Verhaltens, pp. 269–304. *In* Physiological mechanisms in behavior IV. Soc. Exp. Biol. Symp.

Koford, C. B. 1963. Rank of mothers and sons in bands of rhesus monkeys. Science 141:356–357.

Köhler, W. 1925. The mentality of apes. Harcourt, Brace, New York.

Köhler, W. 1947. Gestalt psychology. Liveright Publishing Corp., New York.

Köhler, W., and R. Held. 1949. The cortical correlate of pattern vision. Science 110: 414–419.

Komisaruk, B. R. 1971. Strategies in neuroendocrine neurophysiology. Am. Zool. 11:741–754.

Konishi, M. 1963. The role of auditory feedback in the vocal behavior of the domestic fowl. Z. Tierpsychol. 20:349–367.

Konorski, J. 1948. Conditioned reflexes and neuron organization. Cambridge University Press, New York.

Kortlandt, A. 1940. Eine Uebersicht der angeborenen Verhaltensweisen des Mitteleuropäischen Kormorans (*Phalacrocorax carbo simensis* Shaw and Nodd); ihre Funktion, ontogenetische Entwicklung und phylogenetische Herkunft. Arch. neerl. Zool. 4:401–402.

Kramer, G. 1952. Experiments on bird orientation. Ibis 94:265–285.

Kropotkin, P. 1914. Mutual aid, a factor in evolution. Alfred A. Knopf, Inc., New York.

Kruijt, J. P. 1964. Ontogeny of social behavior in Burmese red junglefowl (*Gallus gallus spadiceua*). Brill, Leiden.

Kruuk, H. 1964. Predators and anti-predator behaviour of the black-headed gull (*Larus ridibundus* L.). Behaviour (Suppl.) XI:1–129.

Kühn, A. 1919. Die Orientierung der Tiere im Raum. Gustav Fischer, Jena.

Kuo, Z. Y. 1932*a*. Ontogeny of embryonic behaviour in Aves, I-II. J. Exp. Psychol. 61:395–430; 62:453–489.

Kuo, Z. Y. 1932*b*. Ontogeny of embryonic behaviour in Aves, IV: The influence of prenatal behaviour on post-natal life. J. Comp. Psychol. 14:109–121.

Kuo, Z. Y. 1938. Further study on the behavior of the cat towards the rat. J. Comp. Psychol. 25:1–8.

Lack, D. 1943. The life of the robin. H. F. & G. Witherby Co., London.

Lack, D. 1954. The natural regulation of animal numbers. Oxford University Press, New York.

Lack, D. 1963. Migration across the southern North Sea studied by radar. Ibis 105 (1):1–54.

Land, M. F. 1972. Stepping movements made by jumping spiders during turns mediated by the lateral eyes. J. Exper. Biol. 57:15–40.

Landsborough-Thompson, A. 1926. Problems of bird migration. H. F. & G. Witherby Co., London.

Lashley, K. S. 1916. The color vision of birds, I: The spectrum of the domestic fowl. J. Anim. Behav. 6:1–26.

Lashley, K. S. 1929. Brain mechanisms and intelligence. University of Chicago Press, Chicago.

Lashley, K. S., K. L. Chow, and J. Semmes. 1951. An examination of the electrical field theory of cerebral integration. Psychol. Rev. 58:123–136.

Lashley, K. S., and J. T. Russel. 1934. The mechanism of vision, XI: A preliminary test of innate organization. J. Genet. Psychol. 45:136–144.

Lehrman, D. S. 1953. A critique of Lorenz's theory of instinctive behavior. Quart. Rev. Biol. 28:337–363.

Lehrman, D. S. 1958a. Induction of broodiness by participation in courtship and nest-building in the ring dove (*Streptopelia risoria*). J. Comp. Physiol. Psychol. 51:32–36.

Lehrman, D. S. 1958b. Effect of female sex hormones on incubation behavior in the ring dove (*Streptopelia risoria*). J. Comp. Physiol. Psychol. 51:142–145.

Lehrman, D. S. 1959. Hormonal responses to external stimuli in birds. Ibis 101:478–496.

Lehrman, D. S. 1961. Hormonal regulation of parental behavior in birds and infra-human mammals, pp. 1268–1382. *In* W. C. Young [ed.], Sex and internal secretions, 3rd ed. The Williams & Wilkins, Co., Baltimore.

Lehrman, D. S. 1963. On the initiation of incubation behaviour in doves. Anim. Behav. 11:433–438.

Lehrman, D. S. 1964. Control of behavior cycles in reproduction, pp. 143–166. *In* W. Etkin [ed.], Social behavior and organization among vertebrates. University of Chicago Press, Chicago.

Lehrman, D. S. 1965. Interaction between internal and external environments in the regulation of the reproductive cycle of the ring dove, pp. 355–380. *In* F. A. Beach [ed.], Sex and behavior. John Wiley & Sons, Inc., New York.

Lehrman, D. S., P. N. Brody, and R. P. Wortis. 1961. The presence of the mate and of nesting material as stimuli for the development of incubation behavior and for gonadotropin secretion in the ring dove (*Streptopelia risoria*). Endocrinology 68:507–516.

Lind, H. 1959. The activitation of an instinct caused by a "transitional action." Behaviour 14:123–135.

Lindauer, M. 1961. Communication among social bees. Harvard University Press, Cambridge, Mass.

Lindsay, W. L. 1880. Mind in the lower animals. D. Appleton & Co., New York.

Lisk, R. D. 1967. Sexual behavior: Hormonal control, pp. 197–239. In L. Martini and W. Ganong [eds.], Neuroendocrinology, 2:197–239.

Lissman, H. W. 1932. Die Umwelt des Kampfisches (Betta splendens Kegan). Z. vergleich. Physiol. 18:65–111.

Lissman, H. W. 1958. On the function and evolution of electric organs in fish. J. Exp. Biol. 35:156–191.

Lluch-Belda, D., L. Irving, and M. Pilson. 1964. Algunas observaciones sobre mamiferos asciones Biol.—Pesqueras. Com. Nacion. Consult. de Pesca y Industrias 10:1–23.

Lockard, R. B. 1971. Reflections on the fall of comparative psychology: Is there a message for us all? Amer. Psych. 26:168–179.

Loeb, J. 1890. Der Heliotropismus der Thiere und seine Uebereinstimmung mit dem Heliotropismus der Pflanzen. Hertz, Würzburg.

Loeb, J. 1906. The dynamics of living matter. Columbia University Press, New York.

Loeb, J. 1918. Forced movements, tropisms, and animal conduct. J. B. Lippincott Co., Philadelphia.

Lorenz, K. Z. 1935. Der Kumpan in der Umwelt des Vogels. J. Ornithol. 83:137–214.

Lorenz, K. Z. 1937. The companion in the bird's world. Auk 54:245–273.

Lorenz, K. Z. 1941. Vergleichende Bewegungsstudien an Anatiden. J. Ornithol. 89: 194–293.

Lorenz, K. Z. 1950. The comparative method of studying innate behaviour patterns, pp. 221–268. In Soc. Exp. Biol. Symp. Physiological mechanisms in animal behavior, Vol. IV, Academic Press, New York.

Lorenz, K. Z. 1952. King Solomon's ring. [Trans. by M. K. Wilson.] Thomas Y. Crowell Company, New York.

Lorenz, K. Z. 1954. Man meets dog. [Trans. by M. K. Wilson.] Methuen & Co., Ltd., London.

Lorenz, K. Z. 1965. Evolution and modification of behavior. University of Chicago Press, Chicago.

Lubbock, J. 1899. On the senses, instincts and intelligence of animals. International Scientific Series, Vol. LXIV. D. Appleton & Co., New York.

MacArthur, R. H. 1959. On the breeding distribution pattern of North American migrant birds. Auk 76:318–325.

McConnell, J. V. 1964. Cannibals, chemicals and contiguity. In Learning and associated phenomena in invertebrates. Anim. Behav. (Suppl.) 1:61–68.

McDougall, W. 1905. Physiological psychology. J. M. Dent & Sons, Ltd., London.

McDougall, W. 1923. Outline of psychology. Charles Scribner's Sons, New York.

McDougall, W. 1938. Fourth report on a Lamarckian experiment. Brit. J. Psychol. 28:321–345, 365–395.

McInnis, Noel. 1969. Gestalt ecology: How do we create our space? In Gwen Barrows [ed.], Fields within fields . . . within fields. The World Institute Council 2 (1):11–23.

Maier, N. R. F., and T. C. Schneirla. 1935. Principles of animal psychology. McGraw-Hill Book Company, New York.

Mainardi, M. 1958. L'insorgenza dell'istinto parentale in un Ibrido sterile fra Columba livia e Columba albitorques. Zoology 4(3):1–7.

Makkink, G. F. 1936. An attempt at an ethogram of the European avocet with ethological and psychological remarks. Ardea 25:1–60.

Manning, A. 1960. The sexual behavior of two sibling Drosophila species. Behaviour 15:123–145.

Manning, A. 1961. Effects of artificial selection for mating speed in Drosophila melanogaster. Anim. Behav. 9:82–92.

Marler, P. 1955. Characteristics of some animal calls. Nature 176:6–8.

Marler, P. 1961a. The logical analysis of animal communication. J. Theoret. Biol. 1:295–317.

Marler, P. 1961b. The filtering of external stimuli during instinctive behavior, pp. 150–166. In W. H. Thorpe and O. L. Zangwill [eds.], Current problems in animal behavior. Cambridge University Press, New York.

Marler, P. 1963. Inheritance and learning in the development of animal vocalizations, pp. 228–243, 794–797. In R. Busnel [ed.], Acoustic behaviour of animals. Elseviere Pub. Co., New York.

Marshall, A. J. 1954. Bowerbirds: Their displays and breeding cycles. Oxford University Press, New York.

Marshall, A. J. 1955. Reproduction in birds: The male. Mem. Soc. Endocrinol. 4: 75–93.

Marshall, F. H. A. 1922. The physiology of reproduction. Longmans, Green & Company, Ltd., London.

Marshall, F. H. A. 1942. Exteroceptive factors in sexual periodicity. Biol. Rev. 17: 68–90.

Mast, S. O. 1938. Factors involved in the process of orientation of lower organisms in light. Biol. Rev. 13:186–224.

Matthews, G. V. T. 1955. Bird navigation. Cambridge University Press, New York.

Maturana, H. R. 1964. Functional organization of the pigeon retina, pp. 170–178. In XXII Int. Physiol. Cong., 1962 Proc. Information processing in the nervous system, Vol. III. Leiden.

Maturana, H. R., and S. Frenk. 1963. Directional movement and edge detectors in the pigeon retina. Science 142:877–979.

Maturana, H. R., J. Y. Lettvin, W. S. McCulloch, and W. H. Pitts. 1960. Anatomy and physiology of vision in the frog. J. Gen. Physiol. 43:127–177.

Meinertzhagen, R. 1950. Note on tameness in birds. Ibis 92:151–152.

Mena, F., and C. E. Grosvenor. 1971. Release of prolactin in rats by exteroceptive stimulation: Sensory stimuli involved. Hormones and Behavior 2:107–116.

Menaker, M. [ed.]. 1971. Biochronometry. Nat'l. Acad. Sci., Washington, D.C.

Merkel, F. W., and W. Wiltschko. 1965. Magnetismus und Richtungs finden Zugunruhigen Rotkelchen (Erithacus rubecula). Vogelwarte 23:71–77.

Messmer, E., and I. Messmer. 1956. Die Entwicklung der Lautäusserungen und einiger Verhaltensweisen der Amsel. Z. Tierpsychol. 13:341–441.

Michael, R. P. 1961. An investigation of the sensitivity of circumscribed neural areas to hormonal stimuli by means of the application of estrogens directly to the brain of the cat, pp. 465–480. In 4th International Neurochemistry Symposium, Proc., Regional neurochemistry. Pergamon Press, New York.

Michael, R. P., and E. B. Keverne. 1968. Pheromones in the communication of sexual status in primates. Nature 218:746–749.

Michael, R. P., and E. B. Keverne. 1970. Primate sex pheromones of vaginal origin. Nature 225:84–85.

Michael, R. P., and E. B. Keverne. 1971. Pheromones: isolation of male sex attractants from a female primate. Science 172:964–966.

Michel, Z. 1928. Les ondes cosmiques et la vie. Rev. Gen. Sci. Pur. Appl. 39:48–52.

Miller, N. E. 1959. Liberalization of basic S-R concepts: extensions to conflict behavior, motivation and social learning, pp. 196–292. In S. Koch [ed.], Psychology: a study of a science, Vol. 2. McGraw-Hill Book Company, New York.

Miller, G. A. et al. 1960. Plans and the structure of behavior. Holt, Rinehart, and Winston, New York.

Milner, P. M. 1960. Learning in neural systems, pp. 190–203. In M. C. Yovits, G. T. Comeron, and G. D. Goldstein [eds.], Self-organizing systems. Pergamon Press, New York.

Mittelstaedt, H. 1962a. Control systems of orientation in insects. Ann. Rev. Entomol. 7:127–198.

Mittelstaedt, H. 1962b. Prey capture in mantids, pp. 51–71. In B. T. Sheer [ed.], Recent advances in invertebrate physiology. University of Oregon Pub., Eugene.

Mittelstaedt, H. 1963. Prey capture in mantids. In B. T. Sheer [ed.], Recent Advances in Invertebrate Physiology 9:51–71. University of Oregon Press, Eugene.

Miyadi, D. 1959. On some new habits and their propagation in Japanese monkey groups, pp. 857–860. In XV International Congress for Zoology, Proc.

Möhres, F. F., 1950. Aus dem Leben unserer Fledermaüse. Kosmos 46:291–295.

Moltz, H. 1963. Imprinting: an epigenetic approach. Psychol. Rev. 70:123–138.

Moltz, H. 1965. Contemporary instinct theory and the fixed action pattern. Psychol. Rev. 72:27–47.

Money, J., and A. A. Ehrhardt. 1971. Fetal hormones and the brain: Effect of sexual dimorphism of behavior, a review. Arch. Sexual Behav. 1:241–262.

Morgan, L. 1896. Habit and instinct. Edward Arnold Publishers, Ltd., London.

Morris, D. 1957. "Typical intensity" and its relation to the problem of ritualization. Behaviour 11:1–12.

Moynihan, M. 1955. Types of hostile display. Auk 72:247–259.

Moynihan, M. 1962. Hostile and sexual behavior patterns of South American and Pacific Laridae. Behaviour (Suppl. VIII).

Müller-Schwarze, D. 1969. Complexity and relative specificity in a mammalian pheromone. Nature 223:525–526.

Muntz, W. R. A. 1962a. Microelectrode recordings from the diencephalon of the frog (rana pipiens) and a blue-sensitive system. J. Neurophysiol. 25:699–711.

Muntz, W. R. A. 1962b. Effectiveness of different colors of light in releasing positive phototactic behavior of frogs and a possible function of the retinal projection to the diencephalon. J. Neurophysiol. 25:712–720.

Muntz, W. R. A. 1963a. The development of phototaxis in the frog (Rana temporaria). J. Exp. Biol. 40:371–379.

Muntz, W. R. A. 1963b. Phototaxis and green rods in urodeles. Nature 199:620.

Muntz, W. R. A. 1964. Vision in frogs. Sci. Amer. 210:110–119.

Murchison, C. [ed.]. 1935. A handbook of social psychology. Clark University Press, Worcester, Mass.

Nelson, K. 1964. The temporal patterning of courtship behavior in the glandulocaudine fishes. Behaviour 24:90–146.

Nelson, K. 1965. The evolution of a pattern of sound production associated with courtship in the characid fish Glandulocauda inequalis. Evolution 18:526–540.

Nelson, K. 1973. Does the holistic analysis of behavior have a future? In P. P. G. Bateson, and P. H. Klopfer [eds.], Perspectives in ethology. Plenum Press, New York. (in press)

Newton, G., and S. Levine. 1968. Early experience and behavior. Charles C Thomas, Publisher, Springfield, Ill.

Nicolai, J. 1964. Der Brutparasitismus der Viduinae als ethologisches Problem. Z. Tierpsychol. 21:129–204.

Nissen, H. W. 1931. A field study of the chimpanzee. Comp. Psychol. Monogr. 8:vi, 122.

Noble, G. K. 1936. Courtship and sexual selection of the flicker (Colaptes auratus luteus). Auk 53:269–282.

Noble, G. K., and A. Schmidt. 1937. Thermal reception by pits in pit vipers and boas. Amer. Phil. Soc., Proc. 77:263–288.

Noble, G. K., and A. Zitrin. 1942. Induction of mating behavior in male and female chicks following injection of sex hormones. Endocrinology 30:327–334.

Nottebohm, F. 1970. Ontogeny of bird song. Science 167:950–956.

Novick, A., and D. R. Griffin. 1961. Laryngeal mechanisms in bats for the production of orientation sounds. J. Exp. Zool. 148:125–146.

Novick, A. 1971. Echolocation in bats: Some aspects of pulse design. Amer. Sci. 59: 198–209.

Olds, J. 1958. Self-stimulation of the brain. Science 127:315–324.

Oppenheim, R. W. 1968. Light responsivity in chick and duck embryos just prior to hatching. Anim. Behav. 16:276–280.

Oppenheim, R. W. 1970. Some aspects of embryonic behavior in the duck (*Anas platyrhynchos*). Anim. Behav. 18:335–352.

Park, O. 1949. Community organization: Periodicity, pp. 528–562. *In* W. C. Allee et al. [eds.], Principles of animal ecology. W. B. Saunders Co., Philadelphia.

Parkes, A. S., and H. M. Bruce. 1961. Olfactory stimuli in mammalian reproduction. Science 134:1049–1054.

Patel, M. D. 1936. The physiology of the formation of the pigeon's milk. Physiol. Zool. 9:129–152.

Patten, W. 1920. The grand strategy of evolution. Richard G. Badger, Boston.

Patterson, I. J. 1965. Timing and spacing of broods in the black-headed gull *Larus ridibundus*. Ibis 107:433–459.

Pavlov, I. P. 1927. Conditioned reflexes: An investigation of the physiological activity of the cerebral cortex. [Transl. G. V. Anrep.] Oxford University Press, London.

Payne, R., and W. H. Drury. 1959. Marksman of the darkness. Nat. Hist. 67:316–323.

Peckham, G., and E. Peckham. 1898. On the instincts and habits of the solitary wasps. Wisc. Geol. Natur. Hist. Surv. Bull. No. 2.

Penfield, W., and L. Roberts. 1959. Speech and brain mechanisms. Princeton University Press, Princeton, N.J.

Pennycuik, C. J. 1960. The physical basis of astro-navigation in birds: Theoretical considerations. J. Exp. Biol. 37(3):573–593.

Petrunkewitsch, A. 1926. Value of instinct as a taxonomic character in spiders. Biol. Bull. Woods Hole 50:427–432.

Pittendrigh, C. S. 1958. Perspectives in the study of biological clocks, pp. 239–268. In Perspectives in marine biology. Scripps Institution of Oceanography, La Jolla, Calif.

Plath, O. E. 1935. Insect societies, pp. 83–141. *In* C. Murchison [ed.], A handbook of social psychology. Clark University Press, Worcester, Mass.

Ploog, D. W. 1964. Verhaltenforshung als Grundlagenwissenschaft für die Psychiatrie. *In* Wiss. Vers. d. Rhein Ver. f. Pschiatrie 123:1–23.

Poulson, H. 1950. Morphological and ethological notes on a hybrid between a domestic duck and a domestic goose. Behaviour 3:99–104.

Premack, D. 1963. Prediction of the comparative reinforcement values of running and drinking. Science 139:1062–1063.

Preyer, W. 1885. Spezielle Physiologie des Embryo. Grieben, Leipzig.

Preyer, W. 1937. Embryonic motility and sensitivity. [Trans. by G. E. Coghill and W. G. Legner.] Monogr. Soc. Res. Child Devel. 11(6):1–115.

Pribram, K. H. 1971. The language of the brain, pp. 316–342. Prentice-Hall, Inc., Englewood Cliffs, N.J.

Pribram, K. H. and F. T. Melges. 1969. Psychophysiological basis of emotion, pp. 316–342. *In* P. J. Vinken and G. W. Brwyn [eds.], Handbook of clinical neurology. Wiley Interscience, New York.

Pulliam, H. R. 1973. Comparative feeding ecology of a tropical grassland finch. Ecology 54:284–299.

Pulliam, H. R., B. K. Gilbert, P. H. Klopfer, D. L. McDonald, L. McDonald, and G. C. Millikan. 1972. On the evolution of sociality, with particular reference to the grassquit, *Tiaris olivacea*. Wil. Bull. 84:77–89.

Ralls, K. 1971. Mammalian scent marking. Science 171:443–449.

Ramsay, A. O., and E. H. Hess. 1954. A laboratory approach to the study of imprinting. Wilson Bull. 66:196–206.

Reiss, R. F. 1964. A theory of resonant networks, pp. 105–137. *In* R. F. Reiss [ed.], Neural theory and modeling. Stanford University Press, Calif.

Renner, M. 1960. The contribution of the honey bee to the study of time-sense and astronomical orientation. Cold Spring Harbor Symp. Quant. Biol. 25:361–367.

Reynaud, G. 1900. The orientation of birds. Bird Lore 2:101–108, 141–147.

Rheingold, H. L. [ed.]. 1963. Maternal behavior in mammals. John Wiley & Sons, Inc., New York.

Richter, C. P. 1945. Further observations on the self-regulatory dietary selections of rats made diabetic by pancreatectomy. Bull. Johns Hopkins Hosp. 76:192–219.

Riddle, O. 1935. Aspects and implications of the hormonal control of the maternal instinct. Amer. Phil. Soc. Proc. 75:521–525.

Riddle, O. 1963. Prolactin or progesterone as key to parental behavior: A review. Anim. Behav. 11:419–432.

Riesen, A. H. 1947. The development of visual perception in man and chimpanzee. Science 106:107–108.

Roeder, K. D. 1963. Nerve cells and insect behavior. Harvard University Press, Cambridge, Mass.

Roeder, K. D. 1971. Communication between bats and moths, pp. 335–343. *In* Topics in the study of life: the bio-source book. Harper & Row, Publishers, New York.

Roeder, K., and A. E. Treat. 1959. Ultrasonic reception by the tympanic organ of nocturnal moths. J. Exp. Zool. 134:127–158.

Romanes, G. J. 1884. Mental evolution in animals. Keegan, Paul, Trench & Co., London.

Romanes, G. J. 1889. Mental evolution in man. D. Appleton & Co., New York.

Rosenblatt, J. S. 1967. Nonhormonal basis of maternal behavior in the rat. Science 156:1512–1514.

Rowan, W. 1925. Relation of light to bird migration and developmental changes. Nature 115:494–495.

Rowan, W. 1931. The riddle of migration. The Williams & Wilkins Co., Baltimore.

Rowell, C. H. F. 1961. Displacement grooming in the chaffinch. Anim. Behav. 9: 38–63.

Rozin, P., and J. W. Kalat. 1971. Specific hungers and poison avoidance as adaptive specializations of learning. Psych. Rev. 78:459–486.

Ruiter, L. de. 1963. The physiology of vertebrate feeding behaviour: Towards a synthesis of the ethological and physiological approaches to problems of behaviour. Z. Tierpsychol. 20:498–516.

Rusinov, U. S. 1956. Electro-physiological research in the dominant area in the higher parts of the CNS, p. 785. In XX Internat. Cong. Physiol. Proc.

Sade, D. S. 1968. Inhibition of son-mother mating among free-ranging rhesus monkeys. Sci. and Psychoanal. 12:18–38.

Sauer, F. 1957. Die Sterneorientierung naechtlich ziehender Grasmuecken. Z. Tierpsychol. 14:29–70.

Schaffner, B. [ed.]. 1954 and 1955. Group processes I and II. Josiah Macy Foundation, New York.

Scharrer, E., and B. Scharrer. 1954. Neurosecretion, pp. 953–1066. In W. von Mollendorf and W. Bargmann [eds.], Handbuch der Microskopischen Anatomie des Menschen. Springer, Berlin.

Schein, M. 1963. On the irreversibility of imprinting. Z. Tierpsychol. 20:462–467.

Schenkel, R. 1948. Ausdrucksstudien an Wolfen. Behaviour 1:81–130.

Schjelderup-Ebbe, T. 1922. Beiträge zur Sozialpsychologie des Haushuhns. Z. Psychol. 88:225–252.

Schleidt, W. M. 1961. Reaktionen von Truthuhnern auf fliegende Raubvögel und Versuche zur Analyse ihrer AAM's. Z. Tierpsychol. 18:534–560.

Schleidt, W. M. 1962. Die historische Entwicklung der Begriffe "Angeborenes auslösendes Schema" und "Angeborener Auslösemechanismus" in der Ethologie. Zeit. f. Tierpsych. 19:697–722.

Schleidt, W. M. 1964a. Ueber die Spontaneität von Erbkoordinationen. Z. Tierpsychol. 21:235–256.

Schleidt, W. M. 1964b. Ueber das Wirkungsfefüge von Balzbewegungen des Truthahnes. Naturwissenshaften 51:445–446.

Schmidt-Koenig, K. 1958. Experimentaelle Einflussnahme auf die 24-Stunden Periodik bei Brieftauben und deren Auswirkungen unter besonderer Beruecksichtigung des Heimkehrvermoegens. Zeit. f. Tierpsych. 15:301–331.

Schmidt-Koenig, K. 1965. Current aspects in bird navigation, pp. 217–278. In D. S. Lehrman et al. [eds.], Advances in the study of behavior. Academic Press, Inc., New York.

Schmidt-Koenig, K., and H. J. Schlichte. 1972. Homing in pigeons with impaired vision. Proc. Nat. Acad. Sci. 69:2446–2447.

Schneiderman, H. A., and L. I. Gilbert. 1964. Control of growth and development in insects. Science 143:325–333.

Schneirla, T. C. 1933. Studies on army ants in Panama. J. Comp. Physiol. Psychol. 15:267–301.

Schneirla, T. C. 1949. Levels in the psychological capacities of animals, pp. 243–286. In R. W. Sellars, V. J. McGill, and M. Farber [eds.], Philosophy for the future. The Macmillan Company, New York.

Schneirla, T. C. 1953. Basic problems in the nature of insect behavior, pp. 656–684. In K. D. Roeder [ed.], Insect physiology. John Wiley & Sons, Inc., New York.

Schneirla, T. C. 1956. Interrelationships of the "innate" and the "acquired" in instinctive behavior, pp. 387–452. In L'instinct dans le comportement des animaux et de l'homme. Masson, Paris.

Schneirla, T. C. 1959. An evolutionary and developmental theory of biphasic processes underlying approach and withdrawal, pp. 1–42. In M. R. Jones [ed.], Current theory and research on motivation, Vol. 7. University of Minnesota Press, Minneapolis.

Schroedinger, Z. 1951. What Is life? Cambridge University Press, New York.

Schultz, F. 1965. Sexuelle Prägung bei Anatiden. Z. Tierpsychol. 22:50–103.

Schutz, F. 1971. Prägung des Sexualverhaltens von Enten und Gänsen durch Sozialeindrücke während der Jugend phase. J. of Neuro-Visceral Relations Suppl. X: 339–357.

Schwartzkopff, J. 1955. On the hearing of birds. Auk 72:340–347.

Schwartzkopff, J. 1962. Vergleichende Physiologie des Gehörs und der Lautäusserungen. Fortschr. Zool. 15:214–336.

Scott, J. P. 1956. The analysis of social organization in animals. Ecology 37:213–221.

Scott, J. P. 1962. Critical periods in behavioral development. Science 138:949–958.

Scott, J. P., and J. L. Fuller. 1965. Genetics and the social behavior of the dog. University of Chicago Press, Chicago.

Sebeok, T. H. [ed.]. 1968. Animal communication. Indiana University Press, Bloomington.

Sewell, G. D. 1970. Ultrasonic communications in rodents. Nature **227**(5256):410.

Shen, S. C. 1953. Cholinesterase in the amphibian nervous system. Yale J. Biol. Med. **26**:172–173.

Sherrington, E. S. 1906. The integrative action of the nervous system. Cambridge University Press, New York.

Simpson, G. G. 1953. The Baldwin effect. Evolution **7**:110–117.

Simpson, G. G., and A. Roe [eds.]. 1958. Behavior and evolution. Yale University Press, New Haven, Conn.

Simpson, M. 1973. The interpretation of agonistic and courtship displays. *In* P. P. G. Bateson and P. H. Klopfer [eds.], Perspectives in ethology. Plenum Press, New York. (in press)

Skinner, B. F. 1938. The behavior of organisms: An experimental analysis. Appleton-Century-Crofts, Inc., New York.

Slater, P. J. B. 1973. Describing sequences of behavior. *In* P. P. G. Bateson and P. H. Klopfer [eds.], Perspectives in ethology. Plenum Press, New York. (in press)

Sluckin, W. 1965. Imprinting and early learning. Aldine Pub. Co., Chicago.

Sluckin, W., and E. A. Salzen. 1961. Imprinting and perceptual learning. Quart. J. Exp. Psychol. **13**:65–77.

Smith, W. J. 1963. Vocal communication of information in birds. Amer. Natur. **97**:117–125.

Sokolov, E. N. 1963. Perception and the conditional reflex. The Macmillan Company, New York.

Sommer, R. 1969. Personal space. Prentice-Hall, Inc., Englewood Cliffs, N.J.

Spalding, D. 1872. Instinct with original observations on young animals. Reprinted *In* Brit. J. Anim. Behav. **2**:2–11.

Spence, K. W. 1960. Behavior theory and learning. Prentice-Hall, Inc., Englewood Cliffs, N.J.

Spencer, H. 1855. Principles of psychology. D. Appleton & Co., New York.

Spencer, H. 1864. First principles. D. Appleton & Co., New York.

Spencer, H. 1896. Principles of psychology, 2nd ed. D. Appleton & Co., New York.

Sperry, R. W. 1956. Experiments on perceptual integration in animals. Psychiat. Res. Rep. **6**:151–160.

Spurway, H. 1956. *In* L'instinct dans le comportement des animaux et de l'homme, pp. 518–519. Masson, Paris.

Starling, E. H. 1905. The chemical correlation of the functions of the body. Lancet **2**:339–341.

Steinbach, E. 1913. Feminierung von Männchen und Maskulierung von Weibchen. Zentralbl. Physiol. **27**:717–723.

Stephens, G. C. 1957. Influence of temperature fluctuations on the diurnal melano-phone rhythm of the fiddler crab, *Uca*. Physiol. Zool. 30:55–69.

Stevens, S. 1951. Handbook of experimental psychology. John Wiley & Sons, Inc., New York.

Stresemann, E. 1951. Die Entwicklung der Ornithologie. Peters, Berlin.

Sutherland, N. S. 1959. Visual discrimination of the shape by *Octopus*: Circles and squares, and circles and triangles. Quart. J. Exp. Psychol. II:24–32.

Sutherland, N. S. 1960. Theories of shape discrimination in *Octopus*. Nature, London 186:848–860.

Szymanski, J. S. 1914. Eine Methode zur Untersuchung der Ruhe und Aktivitaetsperi-oden bei Tieren. Pflueger Arch. Physiol. 158:343–385.

Teitelbaum, P. 1964. Appetite. Amer. Phil. Soc. Proc. 108:464–472.

Teitelbaum, P., and A. N. Epstein. 1962. The lateral hypothalamic syndrome: re-covery of feeding and drinking after lateral hypothalamic lesions. Psychol. Rev. 69:74–90.

Teitelbaum, P., and A. N. Epstein. 1963. The role of taste and smell in the regulation of food and water intake, pp. 347–360. Symp. Olfact. Taste. I Proc. Pergamon, Oxford.

Terkel, J., and J. S. Rosenblatt. 1971. Aspects of nonhormonal maternal behavior in the rat. Hormones and Behavior 2:161–171.

Terkel, J., and J. S. Rosenblatt. 1972. Humeral factors underlying maternal behavior at parturition: Cross-transfusion between freely moving rats. J. Comp. Physiol. Psych. 80:365–371.

Thauzies, A. 1898. L'Orientation. Rev. Sci. Paris 61:392–397.

Thesleff, S. 1962. A neurophysiological speculation concerning learning. Perspect. Biol. Med. 5:293–295.

Thorndike, E. L. 1911. Animal intelligence. The Macmillan Company, New York.

Thorndike, E. L. 1932. Rewards and punishment in animal learning. Comp. Psychol. Monogr. 8(22):1–65.

Thorpe, W. H. 1956. Learning and instinct in animals. Methuen & Co., Ltd., London.

Thorpe, W. H. 1961. Bird-song. Cambridge University Press, New York.

Thorpe, W. H. 1963a. Learning and instinct in animals, new ed. Methuen & Co., Ltd., London.

Thorpe, W. H. 1963b. Ethology and the coding problem in germ cell and brain. Z. Tierpsychol. 20:529–551.

Tinbergen, N. 1951. The study of instinct. Oxford University Press, New York.

Tinbergen, N. 1953a. Social behaviour in animals. Methuen & Co., Ltd., London.

Tinbergen, N. 1953b. The herring gull's world. Collins, London.

Tinbergen, N. 1958. Curious naturalists. Country Life, London.

Tinbergen, N. 1959. Comparative studies of the behaviour of gulls: A progress report. Behaviour 15:1–70.

Tinbergen, N. 1963. Behavior and natural selection. Paper read at Plenary Symposium of the XVI International Congress of Zoology, Washington, D.C., August, 1963.

Tinbergen, N., G. J. Broekhuysen, V. Feekes, J. G. W. Houghton, H. Krunk, and E. Szule. 1962. Egg shell removal by the black-headed gull, *Larus ridibundus* L. a behaviour component of camouflage. Behaviour 19:74–117.

Tinbergen, N., and A. C. Perdeck. 1950. On the stimulus situation releasing the begging response in the newly hatched herring gull chick (*Larus a. argentatus* Ponstopp). Behaviour 3:1–38.

Tolman, C. W. 1964. Social facilitation of feeding behaviour in the domestic chick. Anim. Behav. 12:245–251.

Tolman, C. W., and G. F. Wilson. 1965. Social feeding in domestic chicks. Anim. Behav. 13:134–142.

Tolman, E. C. 1932. Purposive behavior in animals and men. Century Co., New York.

Tolman, E. C. 1959. Principles of purposive behavior, pp. 92–157. *In* S. Koch [ed.], Psychology: a study of a science, Vol. 2. McGraw-Hill Book Company, New York.

Tryon, R. C. 1940. Genetic differences in maze learning in rats, pp. 111–119. *In* National Society for the Study of Education. The 39th Yearbook, I. Public School Publication. Bloomington, Ill.

Tugendhat, B. 1960. The disturbed feeding behavior of the three-spined stickle-back: I. Electric shock is administered in the food area. Behaviour 16:3–4.

Turner, E. R. A. 1964. Social feeding in birds. Behaviour 24:1–46.

Uexküll, J. von. 1909. Umwelt und Innenwelt der Tiere. Springer-Verlag, Berlin.

Verworn, M. 1889. Psycho-physiologische Protistenstudien. Fischer, Jena.

Vince, M. A. 1960. Developmental changes in responsiveness in the great tit (*Parus major*). Behaviour 15:219–243.

Viquier, C. 1882. Le sens de l'orientation et ses organes chez les animaux et chez l'homme. Rev. phil. France létr. 14:1–36.

Waddington, C. H. 1966. Principles of development and differentiation. The Macmillan Company, New York.

Walk, R. D., and E. J. Gibson. 1961. A comparative and analytic study of visual depth perception. Psychol. Monogr. 75:1–44.

Waller, P. F., and M. B. Waller. 1963. Some relationships between early experience and later social behavior in ducklings. Behaviour 20:343–363.

Walsh, G. 1964. Physiology of the nervous system. Longmans, Green & Company, Ltd., London.

Warden, C. J. 1927. The historical development of comparative psychology. Psychol. Rev. 34:57–58, 135–168.

Warner, L. H. 1931. The present status of the problems of orientation and homing by birds. Quart. Rev. Biol. 6:208–214.

Washburn, S. L., and I. deVore. 1961. Social behavior of baboons and early man, pp. 91–105. *In* S. L. Washburn [ed.], Social life of early man. Viking Fund Publications, Chicago.

Watson, J. B. 1913. Psychology as the behaviorist views it. Psychol. Rev. 20:158–177.

Watson, J. B. 1915. Studies on the spectral sensitivities of birds. Pap. Tortugas Lab. Carnegie Inst. Wash. 7:85–104.

Watson, J. B. 1930. Behaviorism. W. W. Norton & Co., Inc., New York.

Watson, J. B., and K. S. Lashley. 1915. An historical and experimental study of homing. Carnegie Inst. Wash. Pub. 211:7–60. (Also as Pap. Tortugas Lab. Carnegie Inst. Wash. 7:7–60.)

Wecker, S. C. 1963. The role of early experience in habitat selection by the prairie deermouse, *Peromyscus maniculatus bairdi*. Ecol. Monogr. 33:307–325.

Weidmann, U. 1956. Observations and experiments on egg-laying in the black-headed gull (*Larus ridibundus* L.) Brit. J. Anim. Behav. 4:150–161.

Weiskrantz, L. 1958. Sensory deprivation and the cat's optic nervous system. Nature 181:1047–1050.

Weisman, A. 1893. The germ-plasm: A theory of heredity. Charles Scribner's Sons, New York.

Weiss, P. A. 1939. Principles of development. University of Chicago Press, Chicago.

Weiss, P. A. 1941. Self-differentiation of the basic patterns of coordination. Comp. Psychol. Monogr. 17:1–96.

Wells, M. J. 1959. Functional evidence for neurone fields representing the individual arms within the central nervous system of *Octopus*. J. Exp. Biol. 36:501–511.

Wells, M. J. 1962. Brain and behaviour in cephalopods. Stanford University Press, Stanford, Calif.

Welsh, J. H. 1938. Diurnal rhythms. Quart. Rev. Biol. 13:123–139.

Wenner, A. M. 1962. Sound production during the waggle dance of the honey bee. Animal Behav. 10:79–95.

Wenner, A. M. 1967. Honey bees: Do they use the distance information contained in their dance maneuver? Science 155:847–849.

Wenner, A. M., P. H. Wells, and D. L. Johnson. 1969. Honey bee recruitment to food sources: Olfactory or language? Science 164:84–85.

Wheeler, R. E. [ed.]. 1967. Hormones and behavior. Van Nostrand Reinhold Company, New York.

Wheeler, W. M. 1923. Social life among the insects. Harcourt, Brace, New York.

Whitman, C. O. 1919. The behavior of pigeons. Carnegie Inst. of Wash. Pub. **257**:1–161.

Whorf, B. L. 1956. Language, thought and reality. John Wiley & Sons, Inc., New York.

Wickler, W. 1961. Ökologie und Stammesgeschichte von Verhaltensweisen. Fortschr. Zool. **31**:303–365.

Wickler, W. 1962. Ei-Attrapen und Maulbrüten bei afrikanischen Cichliden. Z. Tierpsychol. **19**:129–164.

Wiepkema, P. R. 1961. An ethological analysis of the reproductive behavior of the bitterling (*Rhodeus amarus* Bloch). Arch. neerl. Zool. **14**:103–199.

Williams, G. C. 1971. Group selection. Aldine-Atherton, Inc., Chicago.

Wilson, E. O. 1962. Chemical communication among workers of the fire ant, I, II, III. Anim. Behav. **10**:134–147, 148–158, 159–164.

Wilson, E. O. 1971. The insect societies. Belknap Press, Cambridge, Mass.

Wiltschko, W. 1971. About the influence of magnetic total intensity and inclination on directions preferred by migrating European robins *Erithacus rubecula*, pp. 569–578. *In* Animal orientation and navigation. NASA, Washington, D.C.

Witschi, E. 1935. Seasonal sex characters in birds and their hormonal control. Wilson Bull. **47**:177–188.

Wojtusiak, R. J. 1946. Hypothesis of sensibility to infra-red rays as an attempt to explain some problems of orientation in animals, pp. 28–29. *In* Académie polonaise des sciences et des lettres. Compt rendu mensuel des séances de la classe des sciences mathématique et naturelles. Cracovie.

Wolf, E., and G. Zehrren-Wolf. 1937. Flicker and the reactions of bees to flowers. J. Gen. Physiol. **20**:511–518.

Wolfson, A. 1942. Regulation of spring migration in juncos. Condor **44**:237–263.

Wolfson, A. 1945. The role of pituitary, fat deposition and body weight in bird migration. Condor **47**:95–127.

Wolfson, A. 1948. Bird migration and the concept of continental drift. Science **108**:23–30.

Wolfson, A. 1958. Role of light and darkness in the regulation of the annual stimulus for spring migration and reproductive cycles, pp. 758–789. *In* XII International Ornithological Congress. Proc. Helsinki.

Wolfson, A. 1959. The role of light and darkness in the regulation of spring migration and reproductive cycles in birds. *In* Photoperiodism and related phenomena in plants and animals. American Association for the Advancement of Science, Washington, D.C.

Wynne-Edwards, V. C. 1962. Animal disperson in relation to social behavior. Olivere Boyd, Edinburgh.

Yeagley, H. L. 1947. A preliminary study of a physical basis of bird navigation. J. Appl. Physics. **18**:1035–1063.

Yeagley, H. L. 1951. A preliminary study of a physical basis of bird navigation. J. Appl. Phys. **22**:746–760.

Yerkes, R. M. 1943. Chimpanzees. Yale University Press, New Haven, Conn.

Yerkes, R. M., and A. W. Yerkes. 1935. Social behavior in infrahuman primates, pp. 973–1033. *In* C. Murchison [ed.], A handbook of social psychology. Clark University Press, Worcester, Mass.

Young, J. Z. 1961. Learning and discrimination in the octopus. Biol. Rev. **36**:32–96.

Young, J. Z. 1964. A model of the brain. Oxford University Press, New York.

Young, W. C. 1961. The hormones and mating behavior, pp. 1173–1239. *In* W. C. Young [ed.] Sex and internal secretions. The Williams & Wilkins Co., Baltimore.

Young, W. C., 1965. The organization of sexual behavior by hormonal action during the prenatal and larval periods in vertebrates, pp. 89–107. *In* F. A. Beach [ed.], Sex and behavior. John Wiley & Sons, Inc., New York.

Young, W. C., R. W. Goy, and C. H. Phoenix. 1964. Hormones and sexual behavior. Science **143**:212–217.

Zajonc, R. B. 1965. Social facilitation. Science **149**:269–274.

Zippelius, H-M., and W. M. Schleidt, 1956. Ultraschall-Laute bei jungen Mäusen. Die Naturwiss. **43**:502.

Zuckerman, S. 1932. The social life of monkeys and apes. Harcourt, Brace, New York.

AUTHOR INDEX

SUBJECT INDEX

A

action specific energy, 44–46
activation by nervous system, 112–113
aggregations, 152
altruism, 165–166, 201
analogy, 213–215
antithesis, principle of, 10
appetitive behavior, 43
associated habits, 9
attention, 231–233
audition, 142
auditory preferences, 227
Auslöser, 42

B

Baldwin effect, 10
bee language, 137–140
behavior control, 218 et seq.
Behaviorism, 63
biogenic law, 210
bits, 219
brain anatomy, 88–90

brain stimulation, 249–250
Broca's center, 89

C

chemoreception, 143
chronometry, 178 et seq.
color preferences, 225–226
communication, 36–38, 167–168
comparative psychology, 60 et seq.
comparative studies, 207–219
compass orientation, 186
competition, 150
concepts and language, 252
conditional reflex, 63–67
conditioning, instrumental, 67 et seq.
conditioning, operant, 66 et seq.
consciousness, 16–20
consummatory behavior, 43
context, 231
convergence, 213–215
cooperation, 151
cortical localization, 89
critical period, 164, 259–262
culture, 212
cybernetics, 178, 217
cycles, 179–183

ANIMAL INDEX

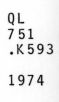